Elements of
STATISTICS

PRENTICE-HALL MATHEMATICS SERIES
Albert A. Bennett, Editor

Elements of
STATISTICS

third edition

Elmer B. Mode

Professor of Mathematics / Boston University

PRENTICE-HALL, INC.
Englewood Cliffs, New Jersey

Third printing......December, 1963

Library of Congress Catalog Card Number 61–10792

Printed in the United States of America
27254C

"The time has come," the Walrus said,
  "To talk of many things:
Of shoes—and ships—and sealing-wax—
  Of cabbages—and kings—"

PREFACE

It has been said that the peculiar characteristic which sets man apart from the other animals is his capacity for capitalizing the achievements of his ancestors. Surely statistics offers a striking example of man's attempt to employ the records of the past in solving the problems of the present and in plotting the course of the future.

Much of the field of elementary statistics has been pretty well organized, so that the general route of progress for the beginner can be mapped out with little difficulty. Yet this route, plainly marked as it is, may be traversed with such facile mathematical skill—and nothing else—as to cause the novice to lose some of the most valuable lessons in simple critical analysis which statistics can teach. There is a peculiar logic associated with the study of statistics. Some call it common sense; others prefer a more exact characterization. In any case, it is difficult to dissociate the purely mathematical aspects from those which are not exactly mathematical but which do require a good brand of clear thinking. It is hoped that this book, dealing as it does with the elementary phases of statistics, will develop on the part of the reader an appropriate attitude of critical reasonableness toward the subject. There is no reason why sound mathematical conclusions should not be accompanied by a healthy appreciation of what such conclusions do *not* assert.

A large amount of important work can be done without extensive preparation in mathematics. Most persons who study elementary statistics are motivated by the fact that it has a direct bearing on a related field of interest:

business, economics, sociology, biology, psychology, education, and others. They want to acquire "sufficient statistical terminology and technique" to enable them to read intelligently and somewhat critically the statistical content of the literature in these fields. If called upon to perform a simple statistical analysis, they wish to be able to handle the fundamental procedures.

The present book favors no particular field of interest. The subjects enumerated above are well represented. There is an unusual abundance and a wide variety of exercises, none of which has been borrowed from other textbooks.

The arrangement of material allows much flexibility in the choice of topics. Many sections and some chapters may be eliminated, if so desired, without affecting the continuity of the book. Those who desire a brief course on methodology can omit such proofs as are present, although these cast valuable light on the concepts and processes involved. In particular, the writer has aimed to describe the full purport of each distinctive statistical concept. A list of references appropriate to an introductory study of this kind will be found at the end of the book.

The present edition, the third, presents a considerable revision of the earlier texts. In a field whose rate of extension and clarification is very rapid, fairly frequent changes in concepts and method of presentation are inevitable. The chief features of this revision are (1) a separate chapter on probability—the cornerstone upon which statistics rests; (2) the exposition of the normal distribution following that of the binomial; (3) a more systematic treatment of the testing of hypotheses; (4) some diminution in the computational aspects of statistics; (5) greater emphasis on the interpretation of results; (6) more up-to-date tables and charts of critical values; and (7) a brief optional chapter on nonparametrics.

The newer expositions have proved to be teachable at the level of the earlier editions of this book, and it has been the author's experience that the more challenging chapters excite the student's interest most of all.

The author is indebted to Sir Ronald A. Fisher and to Messrs. Oliver and Boyd, Ltd., Edinburgh, for permission to reprint Tables III, IV, and Va, from their book, *Statistical Methods for Research Workers*. The author is also grateful to Professor George W. Snedecor and the Iowa State College Press, to Professor Egon S. Pearson and the *Biometrika* Trustees, to Professors Wilfrid J. Dixon and Alexander M. Mood, to Drs. F. Swed and Churchill Eisenhart, and to Professor E. G. Olds, for the use of Tables appearing in the back of the book.

The author's thanks are due his kind colleague Professor Albert Morris, for helpful comments made after reading the manuscript of the first edition, and especially to Professor Albert A. Bennett, whose constructive criticisms and invaluable suggestions added greatly to the interest of the first two editions. In particular the author is deeply in debt to Professor Elizabeth A. Shuhany for her many recommendations for the improvement of the presentation of the material in this, the third edition.

E.B.M.

CONTENTS

1

INTRODUCTION
page 1

2

BASIC MEASURES OF CENTRAL TENDENCY AND VARIABILITY
page 18

3

FREQUENCY DISTRIBUTIONS
page 40

4

THE STATISTICS OF FREQUENCY DISTRIBUTIONS

page 65

5

PROBABILITY

page 87

6

THE BINOMIAL DISTRIBUTION

page 101

7

THE NORMAL
FREQUENCY FUNCTION

page 119

8

INFERENCES
FROM SAMPLE MEANS

page 152

9

DISTRIBUTIONS RELATED TO BINOMIAL
AND NORMAL DISTRIBUTIONS

page 176

10

ESTIMATION

page 190

11

THE CHI-SQUARE
DISTRIBUTION

page 202

12

CURVE FITTING

page 219

13
CORRELATION
page 238

14
THE ANALYSIS OF VARIANCE
page 269

15
NONPARAMETRIC STATISTICS
page 276

REFERENCES
page 288

MATHEMATICAL TABLES
page 291

INDEX
page 313

INTRODUCTION

1

"In 1786, I found that in Germany they were engaged in a
species of political inquiry, to which they had given the name of
Statistics... an inquiry for the purpose of ascertaining the political
strength of a country, or questions respecting matters of state."
SIR JOHN SINCLAIR
The Statistical Account of Scotland, Vol. 21 (1791–1799)

1.1. The Origin of Statistics

It is curious that the modern science of statistics traces its origin to two
quite dissimilar human interests, political states and games of chance. In
the mid-sixteenth century Girolamo Cardano, the Italian mathematician,
physician, and gambler wrote his *Liber de Ludo Aleae* (The Book on Games
of Chance) in which appeared the first known study of the principles of
probability. About a hundred years later, the gambler Chevalier de Méré
proposed to Blaise Pascal the famous "Problem of the Points," which may be
described as follows: Two men are playing a game of chance. The one first
gaining a certain number of points wins the stake. They are forced to quit
before the game is completed. Given the number of points each has won,
how should the stake be divided? This problem offered a real challenge to
the wits of the two astute French mathematicians, Pascal and Fermat. A
lengthy correspondence between the two men led to solutions, not only of
the problems proposed, but of more general ones. The methods employed
by Cardano and Pascal may be said to represent the beginnings of the
mathematics of probability, about which modern statistical theory centers
today. The publication by Laplace in 1812 of the epoch-making *Théorie
Analytique des Probabilités* laid a firm foundation for this theory.

In the mid-eighteenth century, *statistics* itself was born as a word
describing the study of "the political arrangement of the modern states of

1

the known world." The description of states was at first verbal, but the increasing proportion of numerical data in the descriptions gradually gave the new word the quantitative connotation that is associated with it now. From the rather restricted study of data pertaining to a state, statistics branched out into other fields of investigation.

Between 1835 and 1870, the Belgian astronomer Quetelet was applying the theory of probability to anthropological measurements. His conclusions may be summarized and extended by stating that the same general laws of variation governing gambler's luck may be discovered in the statures of soldiers, the intelligence quotients of children, the blood pressures of adults, the speeds of molecules of a gas, and innumerable other aggregates of observations.

In more recent times, an English school of statisticians under the leadership of Karl Pearson (1857–1936) and Ronald A. Fisher, have made notable contributions to both theoretical and applied statistics. The power of general methods based on probability concepts became more clearly perceived, and as a result, applications have been made to many diverse fields of inquiry. An appreciation of the importance of the statistical method in man's attempt to come to grips with a marvelously complex physical and social world is a serious and legitimate aim of any educated person.

1.2. The Meaning of Statistics

The layman frequently conceives of statistics as a mass of figures or a collection of data such as we might find in the publications of the United States Census Bureau, among the records of a school principal, or in the files of a large hospital. The often repeated phrase "Statistics show ... " is likely to imply that a given mass of figures contains salient and unalterable characteristics that can easily be discerned among the mass by any person of normal intelligence. That the word *statistics* may apply to certain aggregates of figures is not to be denied, but that important facts contained therein are easily detected is by no means always true.

A second meaning of *statistics* is simply the plural of *statistic*, where a *statistic* is a certain kind of measure used to evaluate a selected property of the collection of items under investigation. The average weight of a football squad, for example, may be found by adding the weights of the individual players and dividing by their number. The average thus obtained is a *statistic*.

A third meaning of *statistics* is of prime concern to us in this book. It is the science of assembling, analyzing, characterizing, and interpreting collections of data. In this sense, statistics is a field of study, a doctrine concerned with mathematical characterizations of aggregates of items.

Statistics, as a science, is fundamentally a branch of applied mathematics, just as mechanics is mathematics applied to problems connected with bodies subjected to forces. In statistics, the applications may be made to almost any aggregate of observations or measurements. For this reason it is useful in business, economics, sociology, biology, psychology, education, physics, chemistry, agriculture, and related fields.

1.3. Sample and Population

A *population* is a totality of all actual or conceivable objects of a certain class under consideration. More precisely, a population consists of numerical values connected with these objects. Head lengths of criminals, test scores of pupils, thicknesses of washers, lengths of life of electric light bulbs, or numbers of negative replies on a questionnaire may constitute populations of measurements or observations. Such aggregates may be finite or infinite, real or fictitious, but in this book we shall assume all populations to be essentially infinite. Problems connected with finite populations will not be considered here.

A *sample* is a finite number of objects selected from the population. If these are chosen in such a manner that one object has as good a chance of being selected as another, we say that we have a *random sample*. In this connection we should note that some objects may have identical measurements, so that one *measurement* is not necessarily as likely to occur as another.

A group of 10 washers taken at random from a barrel of them constitutes a sample from a larger aggregate or population, from, say 1000 washers, or from the even larger potential product of the machine manufacturing the washers.

If we know the average thickness of the 10 washers, what conclusions can we safely draw concerning the average thickness of all the washers in the barrel? From the statistical characteristics of the sample, what deductions may we make about the number of washers smaller than a desired dimension? This type of problem is basic in statistics and is one to which notable contributions have been made in recent years.

The greatest care must be exercised in selecting a sample that is truly random. The 10 washers should be selected from different parts of the barrel, not from a particular part such as the top. If a sample of student records is to be selected from an alphabetically arranged card file, it would be risky to select all the records from the same portion of the file. Student names might reflect racial factors. A sample of soil should be synthesized from several areas of the garden plot.

In more elaborate analyses, samples must be scientifically constructed so as to include appropriate numbers of different representative groups. Thus, the sample taken in an election poll must contain different economic, geographic, social, racial, or other groups if it is to reflect the voting tendencies of the country as a whole. The problem of obtaining an adequate sample is

Table 1-1 A PAGE OF RANDOM DIGITS

70079	99064	97423	68793	91763	14940	55550	19900	36879	27718
74372	99540	00119	55063	97512	73665	45331	93614	49512	08359
43658	71456	63894	28132	98307	83300	08001	11186	21446	35864
72448	27714	10704	36331	68905	18477	42727	72133	25167	41601
43269	47963	88026	79532	82919	03920	10924	02018	13708	05281
66360	47852	32769	59586	00133	72584	26480	00245	48371	37526
22043	77224	26075	68778	87332	83287	54373	96391	82132	89338
78519	43251	18412	30777	14380	13550	37902	46169	27785	10488
58454	13026	26618	18537	44015	73261	42001	06096	21918	94440
00666	78245	32662	03375	54485	89848	90606	55556	49481	35329
80043	26080	72508	53576	49390	35273	86769	07108	66688	24636
53787	10007	66163	88811	21977	92078	95503	43655	57975	25768
88907	42653	05541	13459	89731	89459	98306	55222	32363	68675
76654	24020	67332	62362	65014	18061	92185	08657	92167	47793
11675	96819	10965	31214	39215	29883	34235	27113	22919	31278
90066	91253	59174	58312	84990	52539	64054	34864	00483	17913
29480	78114	48305	67868	85176	50048	62792	82816	52055	93273
93992	71132	91042	96303	11372	13817	15490	19452	08265	57612
79938	37498	27019	18573	88617	31245	60208	53962	52981	04301
20506	31384	51173	33453	93156	43166	33599	98112	09422	48744
43006	16020	49784	09917	50236	59837	18739	85767	49111	51512
45186	04205	76923	06181	81538	68226	73500	60779	65584	24305
49966	94867	62902	43090	37205	72584	78048	98669	83267	13303
62224	77713	14540	24003	20499	32752	42271	75891	45681	44445
73217	21643	46106	73942	02936	45948	74850	17297	44957	31068
11219	20296	59367	31426	31166	66247	54764	91861	83130	37507
02164	54666	21868	65824	97370	23627	39822	29285	31387	17045
73171	27920	41254	60089	00693	58712	88187	56810	92728	07894
48435	58944	61989	84538	67060	69031	28814	31405	82384	77694
45687	46494	61920	26751	54241	09903	71831	98113	33094	99925
64573	28270	63695	16900	25980	61906	38832	44327	01141	37889
36345	24793	88754	95921	99442	30336	07705	41314	53028	07381
37402	15236	64920	25909	25085	85456	00198	32419	54583	83635
27358	35142	91012	35570	50420	30509	44150	99868	77894	05250
17222	24172	26021	79527	44721	19041	04399	74266	15134	17952
48436	19800	03441	60218	83099	10869	27264	06777	70388	34992
08752	26430	45080	80472	35599	34343	90581	46482	13441	74151
79075	92335	12474	33423	72174	02953	37198	97172	98019	92623
73073	26360	19111	65852	87760	41988	77620	83328	24394	23932
48418	80642	09023	48310	25218	79006	12709	39456	02883	83600
01362	30222	93728	16044	23187	40562	71067	13330	11022	17378
38148	24320	87981	57518	37136	04182	67913	88235	61865	24638
27411	82008	23860	45246	03403	97639	28686	67623	00542	63666
48322	46340	31022	55657	58297	36244	25091	75297	14695	75932
38823	78043	75095	58043	95125	74783	24693	06360	66853	66663
87891	01449	19122	70232	38118	30249	76453	20802	76374	83474
11627	55036	51014	95142	41014	28968	77021	79801	95957	87132
43277	09284	89837	17654	84726	49893	29601	02749	77246	21271
18946	64377	60317	28724	82044	03820	25767	53052	43304	70629
04996	65987	16738	51367	54872	93628	69984	29220	58652	06087

often solved only after careful study and subsequent testing. It is likely to be the major problem in some fields of investigation. Consider, for example, the difficulties involved in the sampling of the fish population of a lake, of the trees of a pine forest, of the red cells of the human blood, and of the voting population of the United States in a presidential election.

1.4. Random Sampling Numbers

Because of the extreme care that must sometimes be exercised in the selection of a sample that is truly random, statisticians frequently employ a "table of random digits." These digits have been produced by a mechanical or electronic process that ensures that each digit obtained is essentially independent of the digits previously obtained and is to be treated as the result of pure chance. A specimen page of random numbers is shown in Table 1-1. One method of use of such a table will be illustrated.

Imagine that a file of college student record cards consists of 1000 cards arranged alphabetically. Let these be numbered 000, 001, 002, 003, ... and so on up to 997, 998, and 999. In order to draw a random sample of 25 cards from this population of 1000 cards we turn to a Table of Random Sampling Numbers, open it at any page and, without looking at this page, place a pencil point among the digits. Assume that the page opened is that shown in the text (Table 1-1), and that the pencil point landed on the twentieth digit in the twenty-ninth row, namely, 8. This is followed, row-wise, by the successive digits 6706069 .... We begin by listing the successive trios of digits 867, 060, 690, 312, 881, 431, 405, 823, and so on, row by row, until 25 different three-digit numbers are obtained. The desired random sample consists of the cards bearing these numbers. In the case of repetitions of a number we merely omit them and continue our listing. More refined methods of selecting the page and initial point are explained in the published tables.

1.5. The Nature of Statistics

The field of statistics is extensive and varied, therefore it may be helpful to describe it in terms of certain dichotomies that characterize it.

Statistical investigations may be *descriptive* or *inferential*. Generally the former type involves fairly simple techniques; the latter demands a somewhat higher order of critical judgment and mathematical methods.

Suppose that we are confronted with a set of measurements or observations actually obtained from life. Such a set usually represents a complex of data from which it is possible to extract an almost unlimited amount of information. For example, the weekly wages of a group of steel workers may yield information of the following kinds: the total wage received by the group, the highest wage, the lowest wage, the most frequent wage, the range of wages, the average wage, the number of wages below $75, the number above $100, the number between $75 and $125, and so on. The task of the statistician is to select a few procedures and measures by means of which the significant aspects of the given data may be thrown into high relief. These aspects may be obtained by means of *classification*, *graphing*, and *averaging*. Because we are concerned only with an effective characterization of the given data themselves as they come to us through observation, and not with estimates or conclusions involving theoretically related populations, we shall distinguish this type of statistical analysis by means of the word *descriptive*. This analysis, confined exclusively to the data before us, deals with methods of recording or tabulating the constituent items, with their visual presentation, with the properties of various kinds of measures, with devices for computing them, and, in fact, with all means of giving a summary description of the data themselves.

The second type of statistical investigation is concerned with conclusions about the population that may be drawn from the data in the sample. If the data are on a large scale they may be treated as practically equivalent to the population and the properties of the sample may be considered to be like those of the population. Sometimes the sample data may constitute the population itself. If the data form a small sample drawn from the population we usually seek to derive the properties of the population from the limited information contained within the sample. It is clear that in any case we must make certain assumptions and interpret our results in the light of them. Theoretical analysis of this type is based upon the mathematical theory of probability. It has important applications, for instance, in industry, medicine, agriculture, education, and the social sciences.

A second dichotomy arises in a consideration of the theoretical and the practical aspects of statistics. *Mathematical* or theoretical statistics seeks to derive the laws which various populations of data and samples derived from them obey, these populations being more or less definitely specified. *Applied* or methodological statistics uses the theoretical results as models for the solution of practical problems.

A third dichotomy occurs in considering the source of the data to be analyzed. If the data are already gathered and you have had no control over them, then you must find out what you can from what you have available. On the other hand, you may consider first the objectives of your investigation and then plan an appropriate experiment from which you can collect significant data. Here you have some control over the kind and amount of data desired. Problems connected with efficiently planned statistical experiments come under the heading of *The Design of Experiments*. In agriculture, industry, and medicine, for example, appropriate designing is highly desirable and indeed, necessary. Large sample methods are most useful when the data are already in; small sample techniques predominate when you gather information from an appropriate experiment.

The data of statistics consist essentially of numerical values derived from measurements, observations, interviews, experiments, and so on—values that, in a given set, are generally different and unpredictable. Such magnitudes are called *random* or *chance variables*, because many chance causes operate to produce them. The outcome of tossing a pair of dice, the score of a student on a test, the length of life of a 40 watt electric light bulb, the per cent of mice dying from the injection of a drug, all are chance variables. More precisely, *if we can assign a definite probability to the event that X will be less than some prescribed value, C, we call X a random variable.* The theory of probability deals with such variables and is the cornerstone of mathematical statistics. The word *variate* is often used to denote a numerical outcome that is the result of chance. Some of the basic definitions and laws of probability are treated in Chapter 5.

1.6. Some Nonmathematical Aspects

Although statistics is fundamentally a mathematical study, there are certain phases of it that may be termed nonmathematical, and that are, in many respects, prerequisite to successful statistical analysis. It is useless to perform elaborate calculations and to derive conclusions from data of questioned reliability and ambiguous meaning.

(1) *Sources.* Data may arise from books, reports, tables, and so forth, or from the results of original investigation, such as interviews, questionnaires, direct observations, actual measurements, and the like. How accurate are the printed sources, and do they measure the precise thing we seek? Did the author collect his data with evident bias? Were the records transcribed accurately? Were the interviews and questionnaires designed to elicit unbiased replies? Were the observations made on an unrepresentative sample?

A check of samples by census tracts in a study of the size of families in a certain area revealed that the samples had indicated larger households than were actually true of the district.... It was then discovered that the enumerators failed to revisit all the families not at home when they called. Obviously the larger the family, the greater the likelihood of someone being home when the enumerator called.[1]

(2) *Definitions.* The nature of the attributes studied in a statistical investigation may profoundly influence the magnitudes involved and the conclusions drawn; for this reason it is essential that basic terms be clearly understood. It is often necessary to make precise definitions, or to describe in detail the method of measurement. If we have records of human head

[1] Jerome B. Cohen, "The Misuse of Statistics," *Journal of the American Statistical Association*, Vol. 33, December, 1938. See also Clyde V. Kiser, "Pitfalls in Sampling for Population Study," in the same *Journal*, Vol. 29, September, 1934.

lengths, we may wish to know the exact points on the head between which the length is taken and the kind of instrument used for measurement. If we are conducting a housing survey and wish accurate information on the number of rooms per dwelling, how shall we define a room? Does the term *room* include Mrs. Gotrox's spacious reception hall, Professor Wise's closet study, Mrs. Smith's unfinished attic, or Mr. Lowe's basement workshop? Much controversy may arise from questions of definition and interpretation.

(3) *Purposes.* The nature and extent of a statistical study are determined largely by the purpose in view. Frequently, the objective is a single concise result displayed in a manner readily understood by "the man on the street." In such cases, finely drawn distinctions, precise mathematical results, and distracting details may be sacrificed to a simple, graphic representation that catches the eye and impresses the mind. Other investigations may aim to establish a thesis—or to destroy one—by means of a comprehensive analysis of all relevant factors. Such studies often require elaborate technical equipment on the part of the reader as well as the writer.

1.7. Misuses of Statistics

Statistics, properly taught, should train the student to read various books and articles containing the results of statistical inquiry with some degree of discrimination. A person of average intelligence may, with moderate instruction, be made aware, at least, of the abuses of the statistical method and of the pitfalls into which impressive figures may lead the layman. Before any systematic discussion of methodology is even begun, it is possible to detect certain types of misleading statements and erroneous conclusions that only too frequently emanate from books, newspapers and periodicals, the platform, and the radio. Let us give a few samples of them; they may help us to detect more subtle errors later.

A New York newspaper in an optimistic editorial cited "statistics on retail sales to prove that conditions were really better than people imagined. Figures showed retail sales in April of 1938 to be higher than sales in April, 1937. The editorial neglected to mention, however, that the Easter holiday in 1937 fell at the end of March and that Easter shopping was therefore concentrated in the middle two weeks of March, whereas in 1938 Easter occurred in the middle of April, resulting, therefore, in a concentration of Easter shopping in the first two weeks in April."[2]

An unpublished report recently sent to the author attempts to prove that the physical condition of the American people is deteriorating. Among the surprising statements made is the following: "In 1900 there was one hospital bed for 240 persons in our country. Today the ratio is better than 1 to 120. Illness is increasing steadily." Obviously, the increase in the number of beds

[2] See the first reference in footnote 1.

is due to the growing realization on the part of the public of the advantages of expert hospital care and to the enlarged demand for such care. We might add that there is a greater social consciousness of the necessity to provide adequate facilities for our sick, and a resulting increase in public and private expenditures for that purpose. The statistics cited hardly reflect the physical condition of the American people.

The air battle of Britain in 1940 was accompanied by a discussion of the method of comparing British and German airplane losses. It was claimed that the majority of German losses were bombers—larger, more expensive machines than fighter planes, and manned by a crew of three or four—whereas the British losses involved largely fighter planes—less expensive to build, and handled by a crew of one. The sheer comparison of numbers of planes lost on each side obviously concealed a substantial amount of significant information.

The frequent misuse of statistics will be further illustrated in the exercises at the end of this chapter.

1.8. Exact and Approximate Numbers

The study of statistics involves the handling of large quantities of numbers, and, in particular, it involves computation with them. The labor need not be arduous if one makes free use of the tables and instruments available as aids in computation, and if one develops an appreciation of the limits of accuracy of the measurements involved. Without going into the theory justifying many rules of computation, it will be the purpose of the remainder of this chapter to discuss briefly important procedures useful in the computational work of statistics.

It is necessary to distinguish between two sources of numbers, those that are directly or indirectly the result of counting, and those that are directly or indirectly the result of measuring, estimating, and so forth. If we count carefully the number of pupils in a classroom or the number of dollars in a purse, we are able to obtain precise answers, such as 46 students or $7.68. No fractions of pupils or of cents are possible. If, for some generous reason, we wish to give $7.68 to each of 46 students, we are able to compute the exact number of dollars and cents required. $46 \times 7.68 = 353.28$, and the exact amount is $353.28. In dealing with numbers such as those described, errors can be avoided by the exercise of ordinary care. Numbers that belong to the type described will hereafter be called *exact numbers*.

Suppose that a surveyor measures the length of a line with a steel tape and records his result as 34.6 feet. This implies, ordinarily, that the true length of the line is between 34.55 and 34.65 feet. Assuming accuracy of the tape and of the work of the surveyor, we can say that, if the line were really a little shorter than 34.55 feet, its length would have been recorded as 34.5 feet;

and, if the line were actually somewhat longer than 34.65 feet, its length would have been recorded as 34.7 feet. The measurement 34.6 feet is approximate and has a maximum possible error of ± 0.05 feet. Thus, the number (of feet) 34.6 is intrinsically of a different character from the number (of dollars) 7.68 mentioned in the preceding section. The former is only an approximate value, and is represented by a numeral of three *significant digits*. Each of the digits 3, 4, and 6 has significance not only by virtue of its character and position, but also by the fact that it expresses definite information about the length of the line. In particular, the last digit, 6, carries with it (in the absence of different information) the implication that the measurement was made to the nearest tenth. The number 7.68 is an absolutely correct value.

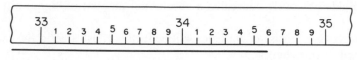

Figure 1–1

It might be denoted by a numeral consisting of an infinite number of significant digits such as 7.6800 ... or 7.67999 .... Numbers that are only approximations will hereafter be called *approximate numbers*. They may arise in two ways: (1) as the result of observation or measurement, or (2) as convenient estimates for numbers that are theoretically exact. Numbers such as π, $\sqrt{2}$, and log 72,348, lead to approximate numbers of the latter kind. Thus, we say that $\pi = 3.1416$ approximately. The number 3.1416 differs from the true value by less than 0.00001. A closer approximation would be 3.14159.

1.9. Significant Digits

Suppose that with a certain measuring instrument we obtain the dimensions of a given rectangle. Let the length found be 41.234 inches and the width 9.432 inches. Suppose, also, that with the same apparatus we find the diameter of a certain steel wire to be 0.027 inches. The number of significant digits in the preceding numerals are 5, 4, and 2, respectively. If there is doubt concerning this statement with regard to 0.027 inches, imagine that *one thousandth of an inch* had been taken as the unit of measurement instead of *one inch*. Then the last value would have been recorded as 27 *thousandths inches*. Thus, recording in a unit 1000 times as large does not affect the significance of the result. The zeros in 0.027 are therefore not counted as significant, for they merely aid in indicating the position of the decimal point.

On the other hand, if the weight of a truckload of stone is set down as 13,500 pounds, it is not known whether this is a numeral of 3, 4, or 5 significant digits. If the weighing apparatus is precise to the nearest *pound*, 13,500 possesses 5 significant digits; if, however, the apparatus weighs only to the nearest *hundredweight*, then 13,500 has but 3 significant digits. The ambiguity in the latter case may be removed by writing the weight as 135 *hundredweight*, or as 135×10^2 *pounds*. The use of powers of 10 is desirable when the magnitudes are very large or very small. As a further example, suppose that the length of the side of a lot of land is measured to the nearest hundredth of a foot and is found to be 106.00 feet. Since the error does not exceed 0.005 feet, the numeral 106.00 has five significant digits. Terminating zeros at the right of the decimal point should be written only when they are significant.

1.10. The Addition and Subtraction of Approximate Numbers

In adding the following approximate numbers:

$$
\begin{array}{r}
13.456 \\
9.30 \\
8.423 \\
913.2 \\
8.21
\end{array}
$$

it would be incorrect to add them as they now stand and to write 952.589 as the sum, for such an addition means that the blank spaces in the columns added have been replaced by zeros when they are actually x's, or unknowns:

$$
\begin{array}{r}
13.456 \\
9.30x \\
8.423 \\
913.2xx \\
8.21x \\
\hline
952.589
\end{array}
$$

Clearly, the sum of a column of digits containing x's is inaccurate. We may state at once the following rule:

The sum or the remainder obtained from the addition or the subtraction of approximate numbers should have no greater precision than that of the least precise number involved in these operations.

For example, to find the sum of the numbers above, we "round off" each numeral to the first decimal place, since the least precise number is the fourth one, 913.2. Thus, we have the following results:

$$
\begin{array}{r}
13.5 \\
9.3 \\
8.4 \\
913.2 \\
8.2 \\
\hline
952.6
\end{array}
$$

Hence, we may write:

Formal answer: 952.6.

Reliable answer: 953.

The possibility of the accumulation of errors due to the "rounding off" process makes the digit in the tenths place of the answer unreliable. In particular, the sum obtained from a long column of numerals may be erroneous in the last two digits of the formal answer for the same reason. Some mathematicians prefer to save all digits in the formal answer despite the fact that they are "semi-reliable."

In rounding off a numeral to a specified decimal place, it is standard practice to round off to the nearest number. Thus, 3.147 rounded off to the second decimal place becomes 3.15. The numeral 7,184,730 rounded off to the nearest ten thousand becomes 7,180,000, or better, 718×10^4. If there are two "nearest" numbers, a good rule is to select the even one. Thus, 0.0345 rounds of to 0.034, and 8.150 rounds off to 8.2.

Subtraction, in these respects, is similar to addition. It should also be noted that the remainder may have fewer significant figures than either the minuend or the subtrahend. Thus,

$$61.4383 - 61.422 = 0.016.$$

In statistics, we are frequently required to add long columns of approximate numbers and to perform simple subtractions. The observance of the simple rules just illustrated and those that are to follow will save much time and effort.

When the average (arithmetic mean) of a small number of items (say, less than 30) is computed, the number of significant figures to be saved is ordinarily the same as that in most of the individual items. Thus, we write the mean of 11.2, 12.4, 9.5, and 10.4 as:

$$\frac{11.2 + 12.4 + 9.5 + 10.4}{4} = 10.9.$$

1.11. The Multiplication and Division of Approximate Numbers

The following working rule is susceptible to formal justification, but it will not be given here.

The product or the quotient obtained from the multiplication or the division of approximate numbers should contain no more significant digits than occur in the least accurate factor.

Thus, if the approximate numbers 41.234 and 9.43 are to be multiplied, the product should contain but three significant digits, the number found in 9.43. If one were to save all digits in this product he would obtain 388.83662. The appropriate product is 389.

Today in America the electrically operated desk calculator is in widespread use in the office, in the laboratory, and often in the classroom. The relative simplicity of many of its operations makes the reduction of numerals to those of fewer digits less urgent. Furthermore, methods of checking results at certain stages of a computation may require the use of more than the justifiable number of digits. In such cases the rounding off of numerals may advantageously be postponed until final answers are obtained.

1.12. Remarks Concerning Approximate Computation

The reader should bear in mind that the rules of Sections 1.10 and 1.11 are only *general* rules and are intended only to lead to simple workable procedures and reasonably accurate results. In other words, these rules are valid *most* of the time, and lead to no serious errors in the type of work to which we are about to apply them. For more detailed information about the limits of error or the precision of results, one must consult larger works on the subject. The subject of approximations, probable error, and the like, cannot be discussed in its entirety within the limits of this book.

1.13. The Slide Rule

Perhaps the most useful single computational device in elementary statistics is the slide rule. This instrument may be described briefly as a mechanical logarithm table. Its chief use in statistics is in multiplication and division, but it may be employed for finding powers, roots, logarithms, and trigonometric functions, and, depending on the type of rule, for other operations. The most common type of slide rule is illustrated in simplified form in Figure 1–2.

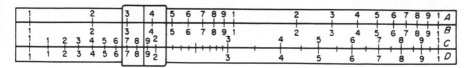

Fig. 1–2. A slide rule (closed).

Details concerning multiplication, its inverse, division, and other operations on the slide rule can be obtained from the instruction book published by the maker of any reputable slide rule. A ten-inch slide rule of standard make is good for at least three-digit accuracy, and this is adequate for most of the work of elementary statistics. The theory is simple, and an hour's practice should make one proficient in multiplication and division. Many other kinds of computation are possible, but are hardly necessary for our work.

1.14. Computation Tables

Tables of powers, roots, and reciprocals will frequently save the computer considerable time and increase the accuracy of his work. Inasmuch as squares, square roots, and reciprocals are required often in statistical work, they may be found in Tables A and B of the Supplementary Tables. In finding square roots from Table A, it is often necessary to make a rough estimate of the answer in order to find the root properly. In this connection we should recall that the pencil method of finding the square root of a number requires the marking off by pairs, in either direction from the decimal point, of the digits composing the numeral. For example, the numeral 341.16785 is marked thus: $3'41'.16'78'5$. The first digit on the left of the square root of 341.16785 is 1, the first approximation to $\sqrt{3}$. The final square root will then have the form:

$$1x.xxxxxx,$$

where the x's are to be determined.

Some of the later work of this book will require the use of special tables of statistical values. These will be discussed at the proper time.

1.15. Bibliography

In a book employing a minimum amount of mathematical equipment, it is obviously impossible to establish rigorously, or even informally, some of the theorems and formulas that are needed. Neither is it feasible to include much interesting and important material concerned with topics of current importance or of special interest in a restricted field. It is hoped that the list of references compiled at the end of this book will prove helpful. These will be referred to by number throughout the book. Other references to special topics will appear occasionally in the text itself.

EXERCISES

Draw, with the aid of the Page of Random Digits, Table 1–1, the following samples. For each sample calculate the average of the sample data.

1. A sample of 10 from Table 3–8.

2. A sample of 20 from the first four columns of heights in Table 3–9.

3. A sample of 25 from the first four columns of weights in Table 3–9.

What is statistically misleading, fallacious, or incomplete in exercises 4 to 8?

4. The climate of Toonerville is particularly agreeable, for the average year-round temperature is 70° F.

5. "During the 15 years before the war, 2385 businesses left Massachusetts, and the migration has not yet halted," declared a well-known business man during a dinner address.

6. A school principal told a certain teacher that her class was below the average for the school. He said that he wanted no classes below the average.

7. Under treatment A, 15 out of 20 patients recovered; under treatment B, only 12 out of 20 recovered. Treatment A is thus shown to be superior to B.

8. During World War II, when the submarine losses of the United States were heavy, a statement was made to the effect that the number of ships sunk was an infinitesimal fraction of the number of voyages.

9. The following statement taken from a reputable publication may be assumed to be correct: "Malignant neoplasms (cancer and allied conditions), which two decades ago were a relatively minor cause of death among children 5 to 14 years of age, now outrank every other disease as a cause of death in this period of life." May one conclude that cancerous conditions are on the increase among children 5–14 years of age? Why?

10. In 1957 the production in the United States increased by 7 per cent; in the Soviet Union the production gain was 17 per cent. Assuming these statements to be correct, what comment can you make on their use for comparison purposes?

11. During the depression years of the 1930's various estimates of unemployment were made from time to time by a number of agencies. In November, 1935, the National Industrial Conference Board estimated the number of unemployed to be 9,177,000; the Labor Research Association estimated it to be 17,029,000. Suggest some reasons for this large discrepancy in the two estimates.

12. In 1942 a newspaper editorial stated, "The 350 strikes in June compare unfavorably with an average of 290 strikes in that month during the peacetime period of 1935–1939." Assuming these figures to be correct what comments are in order?

13. For purposes of census enumeration in the United States, a "farm," in terms of area, has been defined as three or more acres of land upon which agricultural operations are conducted under one management. Places of less than three acres have also been accepted if they had $250 worth of products. Criticize this definition.

    (See R. N. Jessen, "Some Inadequacies of the Federal Censuses of Agriculture," *Journal of the American Statistical Association*, Vol. 44, June, 1949.)

14. From the list of dairies in each of 27 counties of a state, 20 dairies were selected at random. Was the resulting sample of 540 dairies a random one for the state as a whole? Why?

15. What are the objections, if any, to the use of a sample of public opinion obtained by the following methods? (a) Asking questions by telephone. (b) Mailing questionnaires. (c) Interviewing persons on the street. (d) Broadcasting a request for persons to send in their opinions.

16. A ten-gram sample of coal from a carload of soft coal is to be analyzed in the chemical laboratory for ash content. How would you select this sample?

17. Consider the following table of data concerning six-year-old children with caries-free deciduous cuspids, first and second molars, after water fluoridation.

Year	No. children	No. children caries-free	Per cent free
1951	212	30	14.2
1954	193	29	15.0
1957	208	46	22.1

The following statement was made by the authors of this dental investigation. "In 1951, only 14.2 per cent of the six-year-olds were reported with caries-free deciduous cuspids, first and second molars. In 1957 this percentage rose to 22.1, or an improvement of 55 per cent over the 1951 rate." Is this a valid measure of the improvement under fluoridation? Suppose that the 1951 and 1957 per cents free had been 1 and 2 respectively, what would the authors' conclusions have been if the same procedure had been followed? What statement would you make concerning the improvement noted?

18. Of the numbers appearing in the following statements, which are exact and which are approximate?

(a) The shortest man in the group was 5 ft. 6 in. tall.
(b) The maximum temperature for July was 93.8°.
(c) The committee studied the records of 984 students.
(d) His cows produced milk with an average fat content of 5.65 per cent.
(e) Mr. Brown's income for the year was $9,198.68.
(f) The fullback weighed 197 pounds.

19. How many significant digits are there in each of the numerals in the statements below?

(a) 42,000 pounds of fish were landed at Gloucester.
(b) The atomic weight of silver is 107.880.
(c) The speed of light is 300,000 kilometers per second.
(d) The line measured 20.07 feet.
(e) His blood count dropped from 22,000 to 14,000.
(f) The length of the race track at the Olympic games was 100 meters.
(g) The lot of land was 162.00 feet deep.

20. The frontages, in feet, of six adjacent lots on a certain village street were stated to be as follows: 38.97, 147.8, 94.61, 232, 103.04, 56.9. Find their total frontage.

21. Find $(.246)^2 + (.246)^4 + (.246)^6$, where

$$(.246)^2 = .0605;$$
$$(.246)^4 = .00366;$$
$$(.246)^6 = .000222.$$

22. Find the sum of the following approximate numbers:

$$19.37$$
$$1.46$$
$$48.0$$
$$0.93$$
$$205.7$$

23. According to the *World Almanac* for 1940, the imports and exports of the United States for the year 1790 were $23,000,000 and $20,205,156, respectively. The excess of imports if given as $2,794,844. Comment on the precision of this result.

24. In the 1960 presidential campaign a chain of popcorn stands sold popcorn in bags labeled *Kennedy* or *Nixon*, according to the preference of the purchaser. The relative sales of the two kinds of bags were used to estimate voting trends. Criticize this method of polling.

BASIC MEASURES
OF CENTRAL TENDENCY
AND VARIABILITY

2

"Every scheme for the analysis of nature has to face these
two facts, change and endurance."
ALFRED N. WHITEHEAD
Science and the Modern World

2.1. Introduction

Many statistical inferences concerning a population are made from small
numbers of observations. The first step in drawing such inferences consists
in describing the characteristics of the numbers representing a particular set
of observations. The description usually involves averages that summarize
the central tendency of the data and their variability.

An *average* is a typical or representative value. It is a familiar notion
even to a nonmathematical person. In general, he understands the average
bill for electric current to be the sum of the amounts of the bills divided by
the number of the bills. He is likely, also, to understand the average size of
men's shoes sold in a given store to be the size for which there is the greatest
sale. However, these two interpretations of the meaning of "average" are
quite different. They illustrate a frequent mistake made by the "average"
nonmathematical person. The sum of the sizes of all shoes sold in the given
store divided by the number of shoes sold usually yields a fraction, say 8.2,
and this does not correspond, as a rule, to any standard size sold. The "most
popular" size and the "mean" size (the latter) are different concepts. Each

has an important place in statistics; they must not be confused. There are many kinds of averages in statistics. This chapter and Chapter 4 will be devoted to the most important ones.

Yule[1] has described admirably six conditions which a statistical average should, in general, satisfy. They are as follows:

1. The average should be rigidly defined and not left to the mere estimation of the observer.
2. It should be based on all the observations made.
3. It should possess some simple and obvious properties and not be mathematically too abstract.
4. It should be calculated with reasonable ease and rapidity.
5. It should be as stable as possible.
6. It should lend itself readily to algebraic treatment.

Some statistical measures are concerned with the degree of variability of the data. These may describe the dispersion of the variates about a central value, or the complete spread of them. In the sections that follow, certain fundamental measures will be defined and compared.

2.2. A System of Notation

Suppose that the following amounts represent the salaries of the teachers in a certain school:

$6400, 5500, 4600, 6100, 5300, 8700, 4100, 6700, 5700, 5200.

To obtain the average salary known as the *arithmetic mean* we merely add the ten amounts and divide by ten;

$$\tfrac{1}{10}(58{,}300) = 5830.$$

Thus the mean salary is $5830.

In this illustration *salary* is the variable factor which we designate by the letter x. The observed ten salaries constitute the values of the variable or simply the *variates*. Variates are usually represented by the symbol for the variable with subscripts attached. Thus, above, $x_1 = 6400$, $x_2 = 5500$, $x_3 = 4600$, ..., $x_{10} = 5200$. These symbols are read: x sub-one, x sub-two, x sub-three, ..., x sub-ten.

It is necessary at this point to introduce an important notation for variates and other quantities associated with them. Because the calculation of many statistical measures requires the sum of a number of variates, we introduce a standard symbol, the capital Greek letter *sigma*, Σ, to denote a sum, as illustrated below:

$$\sum_{i=1}^{N} x_i = x_1 + x_2 + x_3 + \cdots + x_N.$$

[1] Yule, G. Udny, *An Introduction to the Theory of Statistics*, Charles Griffin, London, 1937, pp. 113–114.

The symbol on the left may be read "summation x_i, i taken from 1 to N," and means that the subscript i is a discrete variable whose values are the integers from 1 to N inclusive. These values are attached to x as subscripts, and the sum of the resulting x's taken. In like manner,

$$\sum_{i=1}^{10} x_i^2 = x_1^2 + x_2^2 + x_3^2 + \cdots + x_{10}^2;$$

$$\sum_{i=1}^{N} x_i y_i = x_1 y_1 + x_2 y_2 + \cdots + x_N y_N;$$

and

$$\sum_{j=1}^{8} f_j(x_j + y_j) = f_1(x_1 + y_1) + f_2(x_2 + y_2) + \cdots + f_8(x_8 + y_8).$$

Note that the variable in the summation is indicated below the summation sign, Σ. In the last three lines, the variable subscripts are i, i, and j, respectively.

There are a few basic theorems involving summations whose proofs follow immediately from the definition of the summation symbol.

THEOREM 2–1. *The summation of the product of a constant and a variable equals the product of the constant and the summation of the variable, that is:*

$$\sum_{i=1}^{N} kx_i = k \sum_{i=1}^{N} x_i.$$

Proof: $\displaystyle\sum_{i=1}^{N} kx_i = kx_1 + kx_2 + kx_3 + \cdots + kx_N$

$$= k(x_1 + x_2 + x_3 + \cdots + x_N)$$

$$= k \sum_{i=1}^{N} x_i.$$

THEOREM 2–2. *The summation of the sum (or difference) of two variables equals the sum (or difference) of their summations, that is:*

$$\sum_{i=1}^{N} (x_i \pm y_i) = \sum_{i=1}^{N} x_i \pm \sum_{i=1}^{N} y_i$$

Proof: $\displaystyle\sum_{i=1}^{N} (x_i \pm y_i) = (x_1 \pm y_1) + (x_2 \pm y_2) + \cdots + (x_N \pm y_N)$

$$= (x_1 + x_2 + \cdots + x_N) \pm (y_1 + y_2 + \cdots + y_N)$$

$$= \sum_{i=1}^{N} x_i \pm \sum_{i=1}^{N} y_i.$$

THEOREM 2–3. *The summation of a constant taken from* **1** *to* **N** *equals the constant multiplied by* **N**, *that is:*

$$\sum_{i=1}^{N} k = Nk.$$

Proof: Since the constant, k, is unaffected by a subscript,

$$\sum_{i=1}^{N} k = k + k + k + \cdots + k$$

$$= Nk.$$

When no ambiguity arises it is common practice to omit the subscripts associated with summation terms. Exercises 23–26 at the end of this chapter are examples of such usage.

2.3. The Arithmetic Mean

The arithmetic mean $\bar{x}$ (read "x-bar") of a set of variates $x_1, x_2, \ldots, x_N$ is defined by the formula,

$$\bar{x} = \frac{1}{N} \sum_{i=1}^{N} x_i. \tag{2.1}$$

This average is not only the most familiar one in use today, but it is also the most basic in theory and the most useful in practice. Unless otherwise stated, the word *mean* customarily denotes the *arithmetic mean*.

The mean of the schoolteachers' salaries calculated at the beginning of the previous section was obtained by Formula (2.1). As a further illustration consider the diameters in centimeters of 13 steel tubes:

2.57, 2.59, 2.64, 2.60, 2.62, 2.57, 2.55, 2.61, 2.50, 2.63, 2.64, 2.56, 2.61.

Their mean diameter

$$\bar{x} = \frac{1}{13}(2.57 + 2.59 + 2.64 + \cdots + 2.61)$$

$$= \frac{1}{13}(33.69)$$

$$= 2.5877.$$

We shall state the mean diameter to be 2.59 cm because of the statement given at the end of Section 1.10.

When the number of items is large, the precision of the mean may exceed somewhat the common precision of the separate items. In fact, the arithmetic mean of N items of common accuracy can be shown to be roughly $\sqrt{N}$ times as accurate as the items themselves.

2.4. A Mathematical Property of the Arithmetic Mean

The *deviation*, d_i, of a variate, x_i, from a mean, $\bar{x}$, is defined by the formula,

$$\mathbf{d}_i = \mathbf{x}_i - \bar{\mathbf{x}}. \tag{2.2}$$

Thus if x_i exceeds $\bar{x}$ the deviation is positive; if x_i is less than $\bar{x}$, the deviation is negative; if $x_i = \bar{x}$ the deviation is zero. The following property involving deviations has important uses in statistics.

THEOREM 2–4. *The sum of the deviations of a set of variates from its arithmetic mean is zero.*

We shall first illustrate this theorem with the data on teachers' salaries (Section 2.2) where $\bar{x} = 5830$.

Table 2–1

x_i	$d_i = x_i - 5830$
6400	570
5500	−330
4600	−1230
6100	270
5300	−530
8700	2870
4100	−1730
6700	870
5700	−130
5200	−630
	4580
Totals	−4580
	0

It is seen from the second column of the table that the sum of the deviations from the mean is exactly zero. (See Exercise 2. 46.)

A formal proof of Theorem 2–4 follows.

$$\sum (x_i - \bar{x}) = \sum x_i - \sum \bar{x} \qquad \text{by Theorem 2–2;}$$

$$= \sum x_i - N\bar{x} \qquad \text{by Theorem 2–3;}$$

$$= \sum x_i - N\left(\frac{1}{N}\sum x_i\right) \qquad \text{by Definition 2.1;}$$

$$= \sum x_i - \sum x_i = 0.$$

2.5. Uses of the Arithmetic Mean

In general, the arithmetic mean is a fairly stable average. It is not unduly affected by a few moderately small or moderately large values, and this stability increases with the total frequency, N. However, one or more extreme values may, at times, profoundly affect its value and render it of doubtful use. The general stability of the mean makes it a highly desirable statistical measure. It has a multitude of uses such as: in meteorology, for obtaining the average temperature or rainfall; in medicine, for discovering the average duration of a disease; in anthropology, for estimating certain average characteristics of a group of human beings; in economics, for computing average wages, prices, index numbers, and so on.

The arithmetic mean is dependent upon the *total* of the variates involved; hence, it is particularly useful in business statistics, as, for example, in averaging sales, production, prices, and so forth, over a specified period. In time series, totals are usually given for the week, month, or year, and from these we may find the average daily, weekly, or monthly figures.

2.6. The Weighted Arithmetic Mean

The *weight* of a variate is a numerical multiplier assigned to it in order to indicate its relative importance.

Suppose that a student receives grades of 88, 81, 78, 74, and 73 in courses carrying credit hours of 2, 3, 4, 3, and 3 hours, respectively. His average grade may be found by multiplying each grade by the number of hours of credit assigned to it. The sum of the weighted values thus obtained is divided by the sum of the weights (the hours of credit), to obtain the weighted mean. Thus,

$$\bar{x} = \frac{(2 \times 88) + (3 \times 81) + (4 \times 78) + (3 \times 74) + (3 \times 73)}{2 + 3 + 4 + 3 + 3}$$
$$= 78.1.$$

The *weighted arithmetic mean* may therefore be defined by the formula:

$$\bar{x} = \frac{\sum_{i=1}^{n} w_i x_i}{\sum_{i=1}^{n} w_i}, \tag{2.3}$$

where w_i is the weight assigned to x_i.

It is often desirable to find the mean of sample means derived from samples of unequal sizes. Thus, if samples of $N_1, N_2, ..., N_k$ have means $\bar{x}_1, \bar{x}_2, ..., \bar{x}_k$ respectively, the *overall* mean of the k samples

$$\bar{x} = \frac{\sum_{i=1}^{k} N_i \bar{x}_i}{\sum_{i=1}^{k} N_i}. \tag{2.4}$$

Here the weights are the sample numbers, N_i. See, for example, Exercise 39.

2.7. The Median

A useful measure of central tendency and one which is easily understood is obtained from that value, if it exists, for which half of the variates are greater and half less. The *median* is the halfway point in a scale of ordered variates.

We may formalize the preceding definition as follows. Let us assume that N variates have been arranged in the following order of magnitude,

$$x_1 \leqq x_2 \leqq x_3 \leqq \cdots \leqq x_N.$$

We then make the following definition:

The median of this ordered set of variates is the value $x_{\frac{N+1}{2}}$ when N is odd,

and the value $\frac{1}{2}(x_{\frac{N}{2}} + x_{\frac{N+2}{2}})$ when N is even.

Let us arrange the diameters of the 13 steel tubes (Section 2–3) in order of magnitude. We obtain,

2.50, 2.55, 2.56, 2.57, 2.57, 2.59, 2.60, 2.61, 2.61, 2.62, 2.63, 2.64, 2.64.

The middle number of this ordered set of 13 is the seventh, 2.60, hence the median diameter is 2.60 cm. There are as many diameters larger than 2.60 as there are smaller.

The salaries of schoolteachers (Section 2.2), arranged in order, are,

4100, 4600, 5200, 5300, 5500, 5700, 6100, 6400, 6700, 8700.

The salary midway between the two middle salaries $5500 and $5700 is $5600.

The usefulness of the median often stems from the fact that one or a few abnormal values that may affect greatly the value of the mean do not disturb the median to an appreciable extent. For example, if the largest salary, above, of a teacher had been $10,700 instead of $8700, the mean would have been raised by $\frac{2000}{10}$ or $200 whereas the median would have remained the same.

In general, the median will have the desirable property that there will be as many variates greater as there are less. As defined, the median is always determinate, but it may not have statistical usefulness. The median becomes useful when there are as many variates greater as there are less.

Consider the numbers of questions answered correctly by 24 students in a multiple-choice examination.

10, 12, 15, 15, 17, 18, 21, 22, 24, 25, 25, 25,
25, 26, 27, 27, 30, 31, 31, 34, 35, 39, 42, 46.

The median of these 24 scores is by definition one half the sum of the 12th and 13th scores, each of which is 25; hence the median is 25 also. But the 10th and 11th scores are also 25, and 25 is not a score such that 12 are smaller and 12 larger. For this reason the use of the median as an average score is open to criticism. It would be better to say that 4 scores were 25, 9 were smaller, and 11 larger. Difficulties of this type arise more often in the case of discrete variates.

2.8. The Range

A simple yet valuable indicator of the variability of a set of observations is the *range*, *R*, defined as the largest variate minus the smallest. Thus for schoolteachers' salaries (Section 2.2),

$$R = 8700 - 4100 = 4600,$$

and for the diameters of tubes (Section 2.3),

$$R = 2.64 - 2.50 = 0.14.$$

The range is not a stable measure because the presence of an unusually large or an unusually small measurement affects it. Nevertheless it has wide application. We may cite the interest of the doctor in the range of the fluctuating temperatures of his patient, of the margin speculator in the "highs" and "lows" of his stock, or of the horticulturist in the climatic extremes of a given geographic region. In industrial quality control (Section 2.13) the ease with which it is understood and the simplicity of its calculation have led to its employment on a large scale.

2.9. The Mean Deviation

A rather natural measure of variation is found by averaging the deviations from the mean, where these deviations are taken without regard to sign. If, in Table 2–1 we change all the minus signs in the second column to plus, we find that the sum of these absolute deviations is 9160. The *mean deviation*,

$$\text{M.D.} = \tfrac{1}{10}\,(9160) = 916,$$

Thus the salaries deviate "on an average" by \$916 from the mean salary of \$5830.

The *absolute value* of a number is its numerical value and is indicated by enclosing the numeral with two vertical bars. For example,

$$|-4| = 4,\ |7| = 7,\ \text{and}\ |5 - 11| = 6.$$

We may now give a formal definition of the mean deviation.

*The **mean deviation, M.D.**, of a set of N variates, $x_1, x_2, x_3, \ldots, x_N$, is defined as the arithmetic mean of their absolute deviations from their arithmetic mean.*

$$\text{M.D.} = \frac{1}{N} \sum_{i=1}^{N} |x_i - \bar{x}|. \tag{2.5}$$

The mean deviation is not an important statistic, but is occasionally used, for example, in laboratory measurements and in tests for the normality of a distribution. (See Section 7.13.) Although it is a sound measure of the dispersion of the data, it is not readily susceptible to algebraic treatment.

2.10. The Variance

In calculating the mean deviation we converted all minus signs to plus before averaging the deviations. Another method of eliminating minus signs is to square the deviations and then average these squares.

The variance, s^2, of a sample of N variates, $x_1, x_2, ..., x_N$, with arithmetic mean, $\bar{x}$, is $1/N$th the sum of the squares of their deviations from the mean. Thus,

$$s^2 = \frac{1}{N} \sum_{i=1}^{N} (x_i - \bar{x})^2. \tag{2.6}[2]$$

The reasons for the use of a *sum of squares* are, (1) it is easily calculated; (2) it may often be broken down into two or more component sums of squares yielding useful statistical information; (3) the minimum property stated in Theorem 2–6 of the next section has practical applications.

Table 2–2

Temperatures x_i	$x_i - \bar{x}$	$(x_i - \bar{x})^2$	x_i^2
47	7	49	2209
38	−2	4	1444
32	−8	64	1024
35	−5	25	1225
41	1	1	1681
39	−1	1	1521
45	5	25	2025
40	0	0	1600
42	2	4	1764
43	3	9	1849
43	3	9	1849
35	−5	25	1225
Totals 480	0	216	19,416

As a simple example of the use of Formula (2.6) let us refer to the first three columns of Table 2–2, which lists the maximum temperatures at 12 selected towns of a certain region on a given date.

The mean temperature is readily found to be $480/12$ or $40°$. The deviations are given in column two, and yield a sum of zero. This affords a check on our work. The squares of the deviations appear in the third column with a sum of 216. Then, $s^2 = \frac{1}{12} (216) = 18.$

[2] Many statisticians prefer to replace N by $N − 1$ in the denominator of Formula (2.6) for the reason stated in Section 10.4.

In most problems the calculation is not as easy as might appear from this example and may be expedited by using the following:

THEOREM 2-5. *The variance of a sample of N variates equals 1/Nth the sum of their squares minus the square of their mean; that is,*

$$s^2 = \frac{1}{N}\sum x^2 - \bar{x}^2. \tag{2.7}$$

Proof:

$$s^2 = \frac{1}{N}\sum (x - \bar{x})^2$$

$$= \frac{1}{N}\sum (x^2 - 2x\bar{x} + \bar{x}^2)$$

$$= \frac{1}{N}\sum x^2 - 2\bar{x}\cdot\frac{1}{N}\sum x + \frac{1}{N}\cdot N\bar{x}^2$$

$$= \frac{1}{N}\sum x^2 - 2\bar{x}^2 + \bar{x}^2$$

$$= \frac{1}{N}\sum x^2 - \bar{x}^2.$$

If we apply Formula (2.7) to the temperature readings in Table 2-2, we may obtain from a table of squares (Table A of the Supplementary Tables) the values shown in the fourth column. Then,

$$s^2 = \frac{1}{12}(19,416) - 40^2$$

$$= 1618 - 1600$$

$$= 18.$$

As another illustration of the use of Formula (2.7), consider the salaries of Section 2.2. The sum of the squares of these ten numbers (salaries) is found, with the aid of Table A to be 354,590,000. The mean was previously found to be 5830. Then

$$s^2 = \frac{1}{10}(354,590,000) - 5830^2$$

$$= 35,459,000 - 33,988,900 = 1,470,100.$$

2.11. Properties of Sums of Squares

Let x_0 be an arbitrary value of a variate. Then

$$\sum (x - x_0)^2 = \sum [(x - \bar{x}) + (\bar{x} - x_0)]^2$$

$$= \sum [(x - \bar{x})^2 + 2(x - \bar{x})(\bar{x} - x_0) + (\bar{x} - x_0)^2]$$

$$= \sum (x - \bar{x})^2 + 2(\bar{x} - x_0)\sum (x - \bar{x}) + \sum (\bar{x} - x_0)^2.$$

But $\Sigma (x - \bar{x}) = 0$, (Theorem 2–4), hence,

$$\Sigma (x - x_0)^2 = \Sigma (x - \bar{x})^2 + \Sigma (\bar{x} - x_0)^2.$$

The last formula illustrates the break-down of a single sum of squares into two sums of squares.

THEOREM 2–6. *The sum of the squares of the deviations is a minimum when taken with respect to the arithmetic mean.*

The proof of this statement follows directly from the preceding formula. The last term $\Sigma (\bar{x} - x_0)^2$ in its right member becomes a minimum, zero, when $x_0 = \bar{x}$; hence, $\Sigma (x_i - x_0)^2$ is a minimum when $x_0 = \bar{x}$.

2.12. The Standard Deviation

The positive square root of the variance is defined as the *standard deviation*. Thus,

$$s = \left[\frac{1}{N} \Sigma (x - \bar{x})^2 \right]^{\frac{1}{2}}. \qquad (2.8)$$

Since the variance of the 12 temperatures of Table 2–2 is 18, their standard deviation is $\sqrt{18} = 4.24$. Similarly the standard deviation for the teachers' salaries is found from the variance (Section 2.10).

$$s = (1{,}470{,}100)^{\frac{1}{2}} = 1212.$$

The standard deviation is, perhaps, the most important and the most widely used measure of variability. A relatively small value of s denotes close clustering about the mean; a relatively large value, wide scattering about the mean. A powerful reason for its usefulness arises from the fact that sums of squares are amenable to simple algebraic manipulation and give rise to useful and interesting relationships. A sum of absolute values, such as occurs in the mean deviation, is not as responsive to mathematical treatment. The standard deviation constitutes a convenient statistical unit for use in the construction of other statistics and in the comparison of these with one another; in other words, many measures are expressed in terms of the standard deviation as a unit.

A good working rule in computing s is to save, at most, one more decimal place than occurs in most of the individual variates. Thus, for the 12 temperatures we could write $s = 4.2°$. In general, however, three digit accuracy is desirable. School teachers' salaries are represented by exact numbers so we may state that their standard deviation is $1212.

2.13. The Relation Between the Standard Deviation and the Range

In certain activities, particularly those of industrial quality control, where the variability of items manufactured in large quantities is under constant inspection, the repeated calculations of standard deviations are time consuming. Moreover, for many employees lacking statistical training, the concept of the standard deviation is difficult to grasp. Intuitively, one

Table 2-3*

Sample no.	1	2	3	4	5	6	7	8	9	10
Weights in grains	38.4	37.4	39.5	37.4	38.0	36.1	38.7	39.5	38.1	39.0
	37.1	37.3	37.4	37.1	39.2	37.6	38.2	39.2	37.8	38.3
	38.6	39.0	38.3	36.5	37.0	38.3	36.2	39.8	36.7	36.9
	38.5	37.7	37.7	36.3	38.2	39.2	38.8	40.8	38.3	38.8
$\bar{x}$	38.15	37.85	38.23	36.83	38.10	37.80	37.98	39.83	37.73	38.25
R	1.5	1.7	2.1	1.1	2.2	3.1	2.6	1.6	1.6	2.1
s	0.62	0.68	0.81	0.45	0.78	1.13	1.06	0.60	0.63	0.82

might expect the standard deviation to increase with the range, and vice versa, and this presumption is generally true. The range is easily understood and quickly computed.

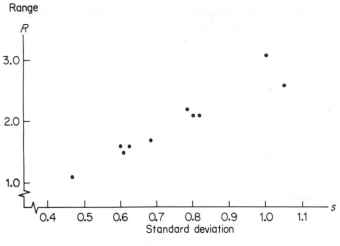

Figure 2-1

The data of Table 2-3 show the weights in grains of 10 samples of 4 explosive charges each. For each of the samples $\bar{x}$, s, and R have been computed. In Figure 2-1 the plot of R versus s shows clearly how one statistic tends to follow the trend of the other. Because of this relationship

* From the American War Standard, *Control Chart Method of Analyzing Data*, Z1.2—1941, published by the American Standards Association, 70 East 45 St., New York 17, N.Y.

estimates can be made of the standard deviation of a population from which a series of small samples have been drawn. Let the small Greek letter, σ (sigma), represent the standard deviation of a population as distinct from s, the standard deviation of a sample.

The outstanding fact for practical purposes is that for samples of a fixed size drawn from a normal population (Chapter 7) the ratio of the mean range, $\bar{R}$, to the standard deviation, σ, tends to be constant. Table 2–4 gives the mean, w, of the ratio $\bar{R}/\sigma$ for convenient values of N.

Table 2–4

TABLE OF THE MEAN OF THE RATIO, $w = \bar{R}/\sigma$, FOR SAMPLES OF N DRAWN FROM A NORMAL POPULATION

N	2	3	4	5	6	7	8	9
w	1.13	1.69	2.06	2.33	2.53	2.70	2.85	2.97

N	10	11	12	13	14	15
w	3.08	3.17	3.26	3.34	3.41	3.47

In practice the R's are computed from small samples, say of 4 or 5, and $\bar{R}$, from 10 to 15 sample R's. Then the value of σ obtained from the table will usually be quite accurate, provided the population is nearly normal.

Example. For the data of Table 2–3, the mean of the ranges, $\bar{R} = 1.96$. Since $N = 4$, we find from Table 2–4 that $w = 2.06$, hence

$$2.06 = 1.96/\sigma$$

or $$\sigma = 0.953.$$

This is an estimate of σ, often symbolized by $\hat{\sigma}$, and comparable to the use of s as discussed in Section 10.4.

2.14. The Appropriateness of an Average

The desirable properties of a statistical average have been summarized in Section 2.1, and the peculiar advantages of each have already been discussed. Nevertheless, it is not always easy to decide which statistical average is most suitable in a given case. The nature of the data and the use to which the average is to be put may influence its selection. Let us illustrate with a few examples.

A certain college is raising endowment funds from its graduates. It is announced that 100 members of the class of 1930 have subscribed a total of $21,400, or an average of $214 per graduate. Suppose that two of these subscriptions were for $5000 and that the remaining 98 ranged from $10 to $250, with most of them in the neighborhood of $100. Obviously, $214 is not typical of the class as a whole. In this case, the median would have been more truly representative, for its value would have been negligibly affected by the two unusually generous subscriptions. The employment of the arithmetic mean has succeeded in doubling approximately the average subscription as measured by the median. Some might argue that the worthy end justifies the mean! At any rate, it is clear that an unrepresentative average may be highly deceptive. We should recall at this point also the fact that solicitors of funds are often faced with the question, "What is the usual subscription?" In such cases, the mode is often a truly typical average.

The health department of a certain town reports annually the median bacteria counts of the milk from dairies selling milk in the town. There are usually twelve counts per year for each dairy. The median is employed as an average in order to minimize the effect of extreme items; for a single, unusually high count might so affect an arithmetic mean count as to create a misleading impression concerning the purity of the milk. An occasional high bacteria count ordinarily is not serious, for the bacteria in the milk of most United States cities are relatively harmless. However, frequent high counts would indicate dirty milk and would be reflected in the median. It is possible to argue that in the case of virulent bacteria, a single high count might lead to fatal results, and, after all, death is permanent. Nevertheless, particularly virulent bacteria are dangerous even in normal counts. It would seem that the median count, on the whole, is as good an average as any, although it does conceal interesting aspects of the data. The range of bacteria counts would supply a valuable additional item of information.

A manufacturer of articles of clothing is not generally interested in mean or median sizes, but in those most frequently demanded. For him, the most typical size is the one most frequently sold, and this average we designate as the *mode*. In like manner, a social worker might be interested in the customary wage of men employed in a certain factory, not the mean or the median wage. The commonest wage might convey certain desired implications. For other purposes, the mean or median wage is usually more suitable.

EXERCISES

In Exercises 1–9, expand the term given.

1. $\displaystyle\sum_{i=1}^{N} F_i$

2. $\displaystyle\sum_{i=1}^{9} f_i$

3. $\displaystyle\sum_{i=1}^{n} f_i y_i$

4. $\displaystyle\sum_{i=1}^{7} x_i^2$

5. $\displaystyle\sum_{i=1}^{n} f_i y_i^3$

6. $\displaystyle\sum_{i=1}^{6} f_i (x_i + y_i)$

7. $\displaystyle\sum_{i=1}^{m} (x_i - x_0)$

8. $\displaystyle\sum_{i=1}^{N} (r_i + s_i)$

9. $\displaystyle\sum_{i=1}^{n} x_i^3$

In Exercises 10–16, write the sums of terms as summations using Σ.

10. $X_1 + X_2 + X_3 + \cdots + X_{10}$

11. $f_1 x_1^2 + f_2 x_2^2 + \cdots + f_n x_n^2$

12. $\dfrac{f_1}{N} + \dfrac{f_2}{N} + \dfrac{f_3}{N} + \cdots + \dfrac{f_n}{N}$

13. $(x_1 - x_0) + (x_2 - x_0) + \cdots + (x_{15} - x_0)$

14. $f_1(y_1 + z_1) + f_2(y_2 + z_2) + \cdots + f_m(y_m + z_m)$

15. $f_1(x_1 - \bar{x})^2 + f_2(x_2 - \bar{x})^2 + \cdots + f_n(x_n - \bar{x})^2$

16. $x_1 y_1 + x_2 y_2 + \cdots + x_{50} y_{50}$

17. Show that $\displaystyle\sum_{x=0}^{5} x^2 = 55$.

In Exercises 18 and 19 write each in a simple form without the use of Σ.

18. $\displaystyle\sum_{i=1}^{10} a.$

19. $\displaystyle\sum_{i=1}^{N} \bar{x}.$

20. Prove that $\displaystyle\sum_{i=1}^{N} (x_i - x_0) - N(\bar{x} - x_0) = 0$.

21. Prove that $\displaystyle\sum_{t=0}^{N} t = \dfrac{N(N+1)}{2}$. (Note the arithmetic progression.)

22. Prove that $\displaystyle\sum_{i=1}^{N} (x_i - \bar{x}) \neq \sum_{i=1}^{N} x_i - \bar{x}$.

Given the pairs of values of x and y in the following tables.

A		B		C		D	
x	y	x	y	x	y	x	y
1	−2	2	3	2	2	2	1
3	0	4	−1	−3	−1	−3	3
5	1	−3	2	5	3	1	−2
6	−1	0	6	0	4	3	0
5	7	5	3			4	1

23. In Table A compute (a) Σx; (b) Σy; (c) Σx^2; (d) Σy^2; (e) Σxy; (f) $\Sigma x \Sigma y$; (g) $\Sigma(x - y)$.

24. In Table B compute (a) Σxy; (b) $(\Sigma y)^2$; (c) Σx^2; (d) $\Sigma x \Sigma y$. Are the values of (a) and (d) equal?

25. In Table C (a) find the value of $\Sigma x \Sigma y - \Sigma x y$. (b) What is the value of the difference between $\Sigma x^2 - \bar{x}^2$ and $\Sigma (x^2 - \bar{x}^2)$?

26. In Table D (a) calculate $\Sigma x \Sigma y - (\Sigma x)^2$; (b) calculate $\Sigma x y + \Sigma y^2$.

27. Prove that $\left(\sum_{i=1}^{N} x_i \right)^2 \neq \sum_{i=1}^{N} x_i^2$.

28. Prove that $\sum_{i=1}^{N} x_i \sum_{i=1}^{N} y_i \neq \sum_{i=1}^{N} x_i y_i$.

29. Assuming that x_1 and x_2 are two unequal positive numbers, prove that the arithmetic mean of their reciprocals (the reciprocal of the *harmonic mean*) is greater than the reciprocal of their arithmetic mean, that is, prove that

$$\frac{1}{2}\left(\frac{1}{x_1} + \frac{1}{x_2}\right) > \frac{1}{\frac{1}{2}(x_1 + x_2)}.$$

30. The tolerances or fatal doses of strophanthus in 0.01 cc per kg for 7 cats were found to be as follows:

$$1.55, \ 1.58, \ 1.71, \ 1.44, \ 1.24, \ 1.89, \ 2.34.$$

Find the mean tolerance.

31. The diameters in mm of 11 pieces of tubing were as follows:

$$6.22, \ 6.25, \ 6.20, \ 6.21, \ 6.20, \ 6.23, \ 6.35, \ 6.23, \ 6.23, \ 6.19, \ 6.26.$$

Compute the mean diameter.

32. The changes in blood pressure of 12 patients after a short period of treatment were as follows:

$$-8, \ 0, \ 12, \ 5, \ -3, \ 0, \ -2, \ 20, \ -12, \ -9, \ -1, \ 3.$$

What was the mean change?

33. The number of prescriptions filled per month at the Pasteur Pharmacy in 1958 were as follows:

Jan	Feb	Mar	Apr	May	June	July	Aug	Sept	Oct	Nov	Dec
341	320	438	363	202	125	117	134	185	229	301	326

Find the mean number filled per month. Assuming 1958 to be a normal year, would this average be a good guide to the number expected to be filled each month in 1959 if that is also assumed to be a normal year? Why?

34. Find the mean grade of a pupil who had 82, 84, and 91 in three 3-credit courses, 77 in a 4-credit course, and 65 in a 2-credit course.

35. Of a graduating class in a certain college, 40% concentrated in Literature or the Arts with a mean grade quotient of 3.12; 25% concentrated in the Natural Sciences or Mathematics with a mean grade quotient of 1.85, and 35% concentrated in the Social Studies with a mean grade quotient of 2.67. What was the mean grade quotient for the entire class?

36. Four different manufacturers supplied the U.S. Government with a certain type of item. Manufacturer A supplied 50% of the items at a cost of $1.35 per item; B supplied 35% at a cost of $1.40; C, 10% at a cost of $1.42, and D, 5% at a cost of $1.47. What was the mean cost per item?

37. In computing the "grade quotient" of a student at a certain college, $A = 4, B = 3, C = 2, D = 1, F = 0$. A plus sign adds 0.3, and a minus sign subtracts 0.3. The numerical equivalents of the grades are weighted according to the number of hours' credit. Compared according to grade quotients, which of the following two records is the better one?

(*a*) B, 3 hrs.; $B-$, 2 hrs.; $C+$, 3 hrs.; D, 3 hrs.; F, 4 hrs.
(*b*) $B-$, 3 hrs.; $C+$, 3 hrs.; $C-$, 4 hrs,; $D+$, 4 hrs.; $D-$, 1 hr.

38. The following data were taken from the *Statistical Abstract of the United States* for 1959.

Grain	Average price (cents per bushel)	Production (1000 bushels)
Wheat	193	950,662
Corn	112	3,422,331
Oats	61	1,300,954
Rye	108	27,243
Barley	88	437,170
Sorghums	97	564,324

Find the average price of grain per bushel according to the following method. Let the price be weighted according to the production; assign a weight of 1 to rye, and compute the corresponding weights for the other grains to the nearest integer.

39. The mean hourly rate of pay for each of four industries and the number employed in each are given below. Compute the overall mean hourly rate.

No. employed	Rate in dollars
463,000	1.30
1,328,000	1.42
217,000	2.68
97,000	2.11

40. The number of cases per year and the mean number of days of in-patients at a certain hospital are tabulated below. What is the overall mean number of days per patient?

Ward	Mean no. days	No. cases
Surgical	20	180
Medical	15	237
Maternity	16	96
Children	14	68

41. In a study of high honors students in economics at Harvard College it was found that in the last thirty years the median income of this group was $11,227 and the arithmetic mean income, $14,664. Why do you suppose these averages differ so much?

42. Fifteen persons made the following contributions to the Red Cross campaign fund:

$5, 7, 2, 15, 2, 6, 12, 150, 5, 5, 20, 10, 12, 10, 8.

Find the mean contribution and the median. Which average do you prefer? Why?

43. The yearly operating expenses of a group of 17 apartment houses follow:

$13,000, 19,000, 17,000, 16,000, 12,000, 27,000, 16,000, 14,000, 13,000, 10,000, 15,000, 13,000, 17,000, 14,000, 16,000, 16,000, 12,000.

Find the mean and the median operating expenses for the group. Which average do you prefer? Why?

44. A group of students were given a special test. The tests were completed in the following numbers of minutes:

11, 15, 16, 19, 22, 24, 28, 30, 35, 38, 40, 52, 58, 70.

Find (a) the mean time and (b) the median time. Which average do you prefer? Why?

45. The temperature of a thermostat was read at two-minute intervals for a period of 20 minutes, with the following results:

Time	0	1	2	3	4	5
Temperature	3.161	3.158	3.159	3.160	3.152	3.162

Time	6	7	8	9	10
Temperature	3.152	3.158	3.162	3.162	3.155

Find the mean temperature and the mean deviation. (From *Journal of Chemical Education*, Feb. 1935.)

46. Consider the diameters of the steel tubes given in Section 2.3. (a) Construct a table like Table 2–1 in order to find the deviations from the mean diameter. The sum of these deviations is not exactly zero. Why? (b) Verify that the mean deviation of the diameters is 0.032 cm.

47. Eight samples of a halibut liver oil product were tested for Vitamin A content. The amounts, x, of Vitamin A measured in 1000 international units per gram, showed the following results:

$$\Sigma x = 187, \quad \Sigma x^2 = 5009.$$

Find the mean and the standard deviation.

48. The diastolic blood pressures of 50 adult males were taken by means of two different types of instruments and the difference, d, between the readings for each individual recorded. It was found that $\Sigma d = 314$ and $\Sigma d^2 = 2336$. Find the mean and the standard deviation of these differences.

49. The anesthetic indexes for 10 dogs were as follows:

2,64, 2.27, 2.63, 2.30, 2.25, 2.25, 2.00, 2.00, 2.40, 2.79.

Show that $\bar{x} = 2.35$ and $s = 0.28$. (The *anesthetic index* is defined as the volume of the agent required to produce respiratory arrest divided by the volume required to produce surgical anesthesia.)

50. Find the mean temperature and the variance for the following temperature readings:

88.1, 88.8, 88.3, 88.5, 88.9, 89.1, 89.4, 89.4, 89.7, 89.8.

51. The numbers of minutes required for a group of 15 pupils to complete a test were as follows:

12, 14, 15, 16, 16, 18, 19, 19, 19, 20, 21, 24, 27, 29, 31.

Find the ratio of the mean deviation to the standard deviation. Carry your calculations to the nearest tenth.

52. Given the following two groups of scores:

A	80	79	70	61	60
B	80	71	70	69	60

Compare (a) the ranges; (b) the means; (c) the standard deviations. (d) What conclusions do you draw?

53. The arithmetic means of the weights of the members of two football squads were 180 pounds for Squad A and 185 pounds for Squad B. The median weights were 181 pounds and 182 pounds respectively. What can you conclude about the relative weights of the members of the two squads?

54. The blood pressures of a group of 60 patients showed $\bar{x} = 140$, $s = 10$, and x_m (median) $= 141$. A second group of 60 showed $\bar{x} = 145$, $s = 13$, and $x_m = 141$. Compare the two groups and give reasons for your statements.

55. List in order of stability the following measures of central tendency: (a) the arithmetic mean; (b) the median; (c) one-half the sum of the highest and the lowest values (the *midrange*); (d) the arithmetic mean of the values left when the lowest 25 per cent and the highest 25 per cent are omitted (the *midmean*).

56. Given the following data for a group of 17 pupils. Compute in each case the measure of central tendency that you consider most useful. Give a reason for each choice.

Heights in inches:

60, 63, 61, 62, 58, 57, 61, 59, 56, 65, 63, 57, 59, 62, 64, 58, 60.

Intelligence quotients:

94, 100, 100, 96, 99, 103, 105, 94, 98, 102, 111, 125, 97, 95, 102, 98, 97.

Scores in a spelling test:

7, 10, 8, 9, 10, 9, 9, 8, 9, 9, 8, 10, 9, 9, 7, 9, 9.

57. Random samples of vitamin capsules taken from two different brands showed the following data for Vitamin C content.

	Brand A	Brand B
No. capsules	100	100
Mean	74	79
Standard deviation	3.6	4.4
Median	75	79

What conclusions do you draw from these results?

58. The mean range of the lengths of collapsible radio antennae for samples of 5 each was found to be 0.020 inches. Estimate the standard deviation. What is the underlying assumption?

59. Refer to the data of Exercise 31. Suppose that three more samples of 11 each were taken and that these showed ranges of 0.08, 0.10, and 0.06. What is the best estimate that you can make of σ?

60. From the data of Table 2–3, compute the ratio, R/s, for each of the 10 samples and note the correspondence among these ratios.

61. The town of Wellesley, Massachusetts, is usually regarded as one of the more prosperous towns of the state. More than 36 per cent of its population reported an income of over $4500 in comparison with the metropolitan area of nearby Boston for which 25 per cent was the corresponding statistic. However, Wellesley showed a median income, $2862, much lower than the metropolitan area. What is the explanation of these seemingly contradictory facts?

62. The mean price per drug prescription by geographical sections is given in the following table (Lilly Digest, 1956). Find the mean and the standard deviation of the prices for the entire country.

Section	No. pharmacies	Average price
New England	61	$2.43
Middle Atlantic	130	2.52
East North Central	249	2.78
West North Central	194	2.59
South Atlantic	93	2.33
East South Central	65	2.38
West South Central	82	2.50
Mountain	84	2.88
Pacific	130	2.90

63. *The harmonic mean of a set of positive variates, x_1, x_2, x_3, ..., x_N, is defined to be the number, x_h, whose reciprocal, $1/x_h$, is the arithmetic mean of $1/x_1$, $1/x_2$, $1/x_3$, ..., $1/x_N$, the reciprocals of the variates.*

$$\frac{1}{x_h} = \frac{1}{N} \sum_{i=1}^{N} \frac{1}{x_i}.$$

A man traveled by automobile 100 miles at 20 mph, 100 miles at 40 mph, and 100 miles at 25 mph. Use the harmonic mean to find his average speed for the 300 miles.

64. Do the preceding problem by means of a weighted mean. Note that the harmonic mean is a disguised weighted mean where the weights are inversely proportional to the variates.

65. Assume the average retail price of bread per loaf in three successive years to be 10, 12, and 20 cents. Compute the average number of loaves that can be purchased for a dollar over these three years. Use both the unweighted arithmetic mean and the harmonic mean of the bread prices in obtaining your answer, and compare the results.

66. *The geometric mean, x_g, of a set of N positive variates, x_1, x_2, x_3, ..., x_N, is defined as the positive Nth root of their product.*

$$x_g = (x_1 x_2 x_3 \cdots x_N)^{1/N}.$$

Find the geometric mean of the numbers 3, 8, and 9.

67. If a variable changes in value from x_1 to x_2, let us define $x_2 - x_1$ as the *absolute* change and x_2/x_1 as the *relative* change. The geometric mean (Exercise 66) is used when relative changes among the variates are more significant than absolute changes. If the price of an article changes from $2 to $1 while the price of a second article changes from $1 to $2, (a) what is the geometric mean and (b) what is the arithmetic mean of the two relative changes? Note that the second average gives an upward bias to the mean of the relative changes in price.

68. If r_i is the ratio of an amount, A_i, in the ith year to the amount, A_{i-1}, in the previous year so that $A_i = A_{i-1}r_i$, then the amount, A_n, at the end of n years is given by the formula

$$A_n = A_0 r_1 r_2 \cdots r_n$$

where $A_1 = A_0 r_1$, $A_2 = A_1 r_2 = A_0 r_1 r_2 \cdots A_n = A_{n-1}r_n$.

If we replace each r_i by an average ratio, r, then

$$A_n = A_0 r^n,$$

where
$$r^n = r_1 r_2 \cdots r_n = A_n/A_0,$$

or
$$r = (r_1 r_2 \cdots r_n)^{1/n}$$

$$= \left(\frac{A_n}{A_0}\right)^{1/n}$$

The average, r, is an example of a geometric mean.

In 1890 there were 62 divorces per 1,000 marriages, and in 1932 there were 163. The increase has been fairly steady. What was the average yearly rate of increase expressed to the nearest tenth of a per cent?

69. An *index number* measures the relative change in the magnitude of a group of related variables. It is the result of attempts to measure changes in a variable that is itself a complex of interdependent variables. For example the "cost of food" is a variable depending upon many other variables such as the prices of flour, beef, pork, potatoes, milk, eggs, and so forth. Index numbers may measure changes in commodity prices, wages, volume of trade, employment, stock quotations, production, and so on.

The index numbers for factory employment in Massachusetts for seven years follow. Find the average yearly rate of decrease of these indexes.

1926	1927	1928	1929	1930	1931	1932
102.0	97.9	91.6	94.3	81.5	73.5	60.6

70. In a congressional investigation made in 1951 of the base salaries of major league baseball players, it was stated that the salaries ranged from $5,000 to $90,000 with $11,000 as the median salary. What conclusions do you draw from these data?

FREQUENCY DISTRIBUTIONS

3

"Figures may not lie, but statistics compiled unscientifically and analyzed incompetently are almost sure to be misleading, and when this condition is unnecessarily chronic the so-called statisticians may well be called liars."

E. B. WILSON
Bulletin of the American Mathematical Society, Vol. 18 (1912)

3.1 Tabulation

Statistical data often come to us as large samples and in a form unsuitable for immediate interpretation. It is usually necessary to group the data into appropriate classes before their general characteristics can be detected and measured. Thus, in Table 3–1 no trend or pattern in the bills for electricity is evident, other than the obvious tendency for most of the amounts to lie between $6 and $7.

Table 3–1
MONTHLY BILLS FOR ELECTRIC CURRENT (*Author's Data*)

5.82	6.84	5.25	6.45	7.26	6.45
5.67	6.09	8.99	6.72	6.46	5.13
3.96	6.42	6.21	7.67	7.40	3.27
4.41	6.00	7.02	6.92	7.80	5.85
4.83	5.25	5.76	6.20	7.40	7.29
5.40	5.76	6.93	6.53	6.11	6.06
6.30	6.27	6.00	6.95	6.91	5.28
6.54	4.77	6.33	4.85	6.67	7.32
5.88	6.72	6.48	6.98	8.59	6.03
6.39	8.55	5.76	5.27	6.81	7.89
7.20	6.24	5.19	6.96	5.85	7.71

Examination of the data shows that the smallest bill is $3.27 and the largest $8.99, so that the *range* of the values is $8.99 minus $3.27, or $5.72. We may conveniently divide this range into 12 *class intervals*, each of length $0.50, as indicated in Table 3–2, and assign each monthly bill to its appropriate class. For purposes of tabulation, the *limits* for the class intervals are set down in order of magnitude in one column and a tally mark is placed in the adjoining column as each amount is assigned to its proper class. When all have been assigned, the tallies are counted and the resulting *frequency* for each class is written in the third column. The completely tabulated record, consisting of a series of class intervals and their associated frequency numbers, is called a *frequency table*. The frequency of a class is merely the

Table 3–2

FREQUENCY TABLE FOR MONTHLY BILLS FOR ELECTRICITY

Class limits	Tallies	Frequency
3.00–3.49	/	1
3.50–3.99	/	1
4.00–4.49	/	1
4.50–4.99	///	3
5.00–5.49	ꟼꟼ //	7
5.50–5.99	ꟼꟼ ///	8
6.00–6.49	ꟼꟼ ꟼꟼ ꟼꟼ ///	18
6.50–6.99	ꟼꟼ ꟼꟼ ///	13
7.00–7.49	ꟼꟼ //	7
7.50–7.99	////	4
8.00–8.49		0
8.50–8.99	///	3

Total Frequency 66

number in that class. Thus, the frequency of the class 6.00–6.49 is 18, which means that there were 18 bills having amounts from $6.00 to $6.49 inclusive. Likewise, there were 4 bills having amounts from $7.50 to $7.99 inclusive, and so on. When data have been classified in such a way that they may be described in terms of class frequencies, the result is called a *frequency distribution*. The *relative frequency* of a class is the fraction obtained by dividing the class frequency by the total frequency. Thus, the relative frequency of electric bills ranging from $5.50 to $5.99 inclusive is $\frac{8}{66}$, or approximately 0.12. This is equivalent to saying that 12 per cent of the bills lie within the interval 5.50–5.99. The frequency distribution is an important initial concept in statistics, for its formation is usually the first step in many statistical investigations.

As a first simple result of the tabulation of the data in Table 3–2, we may say that the distribution is somewhat symmetrical, with the largest frequencies near the center of the table and the frequencies diminishing in either direction from the center. Bills from $6.00 to $6.49 are most frequent, and the average bill probably lies within or near that interval. As we proceed with our study of frequency distributions, we shall notice that distributions which are approximately symmetrical or which are somewhat asymmetrical or *skew* with frequencies diminishing as they approach the ends of the distribution, are the most common types.

3.2. The Frequency Distribution

Consider the frequency distribution of Table 3–3. The measurements of head lengths were apparently made to the nearest millimeter, so that any reading—184 mm, for example—is an approximate number and represents a true length not less than 183.5 mm and not more than 184.5 mm. (See Section 1.8.) The class intervals chosen for these data were 4 mm in width.

Table 3–3

HEAD LENGTHS IN MILLIMETERS OF 462 ENGLISH CRIMINALS, AGE 25–30
(Data from Charles Goring, *The English Convict*, H.M.S. Office, 1913, p. 54)*

(1) Class limits	(2) Class boundaries	(3) Class mark (mid-value)	(4) Frequency
172–175	171.5–175.5	173.5	3
176–179	175.5–179.5	177.5	9
180–183	179.5–183.5	181.5	29
184–187	183.5–187.5	185.5	76
188–191	187.5–191.5	189.5	104
192–195	191.5–195.5	193.5	110
196–199	195.5–199.5	197.5	88
200–203	199.5–203.5	201.5	30
204–207	203.5–207.5	205.5	6
208–211	207.5–211.5	209.5	4
212–215	211.5–215.5	213.5	2
216–219	215.5–219.5	217.5	1

Total 462

* By permission of The Controller of Her Britannic Majesty's Stationery Office.

The class limits (Column 1) are not the same as the *class boundaries* (Column 2), for the class with limits 172–175, for example, may contain lengths as small as 171.5 (the lower boundary) and as large as 175.5 (the upper boundary). The *class mark* of each class interval (Column 3) is the mid-value and has important uses to be discussed later. In most practical work, it is not necessary to construct a frequency table containing all the different columns, 1 to 3, that appear in Table 3–3. Usually, either Column 1, 2, or 3, together with the column of frequencies, are sufficient to specify adequately the classification of the data. It should be noted also that the width of the class interval may be obtained by subtracting any number in any of Columns 1, 2, and 3 from the number just below it. It is *not* necessarily the difference between the limits for a given class. This difference in Table 3–3, is 3, which is not the class interval.

As an illustration of data of a slightly different kind, consider that of Table 3–4. There is no question of choice of class interval here, for the classification is a natural one. Class limits, boundaries, and mid-values have no real meaning here, although later we may have occasion to establish artificial boundaries for such data. (Section 7.10.)

Observations occurring in statistical work are often recorded with scientific care. Thus, the head lengths of criminals were measured "to the nearest" millimeter so that the class boundaries or end-values could be established with precision. On the other hand, common commercial methods of

Table 3–4
FREQUENCY DISTRIBUTION FOR NUMBERS OF HEADS WHICH
APPEARED WHEN SIX PENNIES WERE TOSSED 128 TIMES

No. of heads	Frequency
0	2
1	10
2	28
3	44
4	30
5	13
6	1

Total 128

describing classes contain such phrases as "under ten years of age," "up to fifty pounds," and so forth. Inasmuch as ages are usually recorded only to the whole completed year, a person aged 9 years, 11 months, for example, would be *nine years old*, although his nearest birthday is the tenth. Occasionally, ages *are* recorded to the nearest birthday. An express package weighing 49 pounds and 15 ounces should fall within the category "up to 50 pounds," although an inaccurate weighing machine or a careless weigher might record it as exactly 50 pounds. When the classification "from... up to..." or a similar one is employed, as it often is for the census, taxes, and other data, it includes the lower end-value but not the upper.

A classification according to age, such as that of Table 3–5, will be understood in this book to include values exactly equal to the lower limit and up to but not including the lower limit of the next class above. For example, the age class 10–14 of Table 3–5 means ages of exactly 10 years and up to and including 14 years and any fraction thereof, but not 15 years. Note that this means that the mid-values or class marks are 2.5, 7.5, 12.5, 17.5, and so forth and *not* 2, 7, 12, and so forth. It is perhaps unfortunate that the word *limit* as used here is different from the customary use in the field of mathematics. The class limits of ages just mentioned, 5, 9, 10, 14, and so on, are not constants approached by a certain variable. They are convenient values used to designate the extremities or approximate extremities of certain ranges.

Frequency distributions often come to us "ready made," in many different forms, and it is sometimes difficult, if not impossible, to decide what the original basis of tabulation was. A group of men might have been weighed to the nearest half-pound or to the nearest quarter-pound; a set of discount rates at the bank might have been expressed to the nearest sixteenth of a per cent or to the nearest hundredth; ages might have been given to the nearest birthday or to the last birthday. When the statistician meets such already classified distributions, he must usually exert good judgment in deciding the probable basis of classification. Sometimes he must trace the original data.

The *class mark* is a convenient value used to represent or typify the values within a class interval. In this book it will be the mid-value of the interval. Inasmuch as the purpose of many statistical investigations is to present data in a succinct form, readily understood by the nonmathematical reader, the use of simple class limits and class marks is desirable. It is not uncommon to find that an attempt to achieve unerring precision has obscured simple but important facts and has led only to mathematical pedantry.

3.3. Class Intervals

When a certain characteristic of a large number of individuals is recorded quantitatively, the set of values obtained belongs either to a *continuous* set or to a *discrete* set of numbers. Thus, the head lengths of criminals or the average daily July temperatures of a certain city belong to continuous sets, since any head length or any temperature (between reasonable limits) is possible. On the other hand, the numbers of heads appearing when six pennies are tossed or the sizes of shoes sold in a given store belong to discrete sets, since intervals or gaps exist in the range of the data, within which no values exist. Two and one-half heads cannot appear when six pennies are tossed, nor can a shoe of size 6.83 be purchased at a store.

Given data that are to be assembled into a frequency distribution, the choice of an appropriate class interval is guided by a few simple rules. In the first place, grouping is generally considered disadvantageous when the total frequency is below 50. When grouping is desirable, the number of classes should, if possible, range from about 10 to 25. Fewer than ten classes leads to inaccuracies in computation, and more than thirty classes usually leads to cumbersome operations. Integral values, unity especially, constitute frequent class intervals in the case of discrete sets. Simple fractions, such as one-half in the case of shoe sizes, are also common. In both continuous and discrete data, it is good practice, when feasible, to express the class limits in terms of integers or simple fractions. These rules follow from the desire to make all graphical representations and computations as simple and as accurate as possible.

3.4. Graphical Representation

The frequency distribution of Table 3–3 may be pictured by means of a *histogram* (Figure 3–1). This is constructed as follows. The twelve class intervals of four millimeters each are laid off to a convenient scale on a

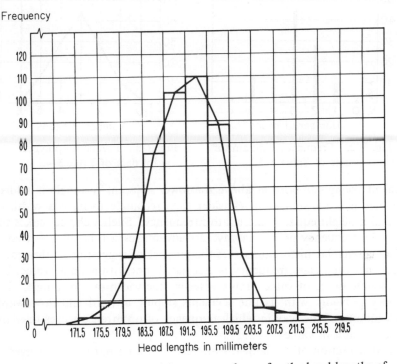

Fig. 3–1.   Histogram and frequency polygon for the head lengths of 462 English criminals.

horizontal line. The end-points of these intervals correspond to the class boundaries. The corresponding class frequencies are measured to a convenient scale on a vertical line, and a rectangle is constructed for each class. Thus, the histogram consists of a set of adjacent rectangles, the bases of which equal the class width and the altitudes of which equal the corresponding class frequencies. The total area of the histogram equals the sum of the areas of the rectangles; and in the usual case, where the class interval or width is the same, this total area equals the product of the common width by the sum of the frequencies (the heights). If we call the class width k, and the sum of the frequencies N, the area of the histogram will be Nk.

By joining the adjacent mid-points of the upper bases with line segments, as indicated in Figure 3–1, a *frequency polygon* is obtained. When the polygon is continued to the horizontal axis just outside the range of lengths, as in the figure, the total area under the polygon will be equal to that of the histogram. For each triangular portion of a rectangle cut off by the polygon, an

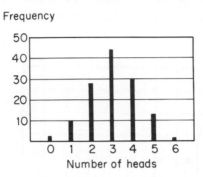

Fig. 3–2. Frequency diagram for
data of Table 3–4.

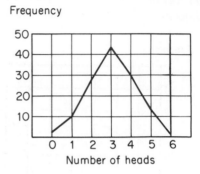

Fig. 3–3. Frequency polygon
for data of Table 3–4.

equivalent triangular portion above the lower adjacent rectangle is added. Sometimes it is convenient to consider each rectangle to have a width of unity and a height equal to the relative frequency. Such a choice makes the area of the frequency polygon unity. Its advantages will be apparent later.

The rectangular histogram serves the purpose of suggesting to the reader the fact that each class frequency is associated, not with a single value, but with a whole range of values, represented by the width of a rectangle. The frequency polygon, with its line segments sloping upward and downward, gives a picture of the way in which frequency of occurrence varies over the complete gamut of values.

The distribution of Table 3–4 is discrete, and does not involve ranges of values. For this reason, many statisticians prefer to use heavy vertical lines instead of vertical rectangles in the frequency diagram. (See Figure 3–2.) There are, however, certain advantages in using a polygon representation, as in Figure 3–3. For reasons to be discussed later, one may also wish to use the histogram.

The monthly bills for electric current (Table 3–2) form a frequency distribution slightly different from the two distributions just discussed. Strictly speaking, the data are discrete inasmuch as the amounts do not involve portions of a cent. The class limits are also the class boundaries, but the upper boundary of any class is not the lower boundary of the next class, as it is in the case of the distribution of head lengths. Nevertheless, in constructing the histogram, one may, for simplicity, disregard this fact and use the boundaries shown in Figure 3–4. These are more desirable for the purpose of popular presentation.

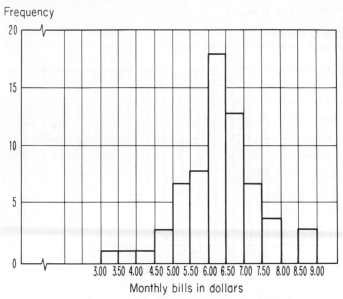

Fig. 3–4. Histogram for data of Table 3–2.

Data are often tabulated with unequal class intervals, as in Table 3–5. In constructing the histogram for such data, one must keep in mind the fact that the areas of the rectangles should be proportional to the corresponding frequencies; hence, if some rectangles have *double* the width of others, the

Table 3–5

DEATHS FROM TYPHOID FEVER, BY AGE, IN THE UNITED STATES, 1930
(*Mortality Statistics*, 1930, U.S. Dept. of Commerce, p. 266)

Age	No. of deaths
0– 4	270
5– 9	425
10–14	616
15–19	939
20–24	881
25–29	579
30–34	381
35–44	591
45–54	456
55–64	275
65–74	141
75 and over	45

altitudes of the former must correspond to *half* the frequencies of the larger class. Thus, in Figure 3–5, the height of the rectangle having the base 35–45 corresponds to 295.5 (half the frequency 591), but the area of this rectangle corresponds to 591. Again, for the purpose of simplicity, we use round numbers on the age scale.

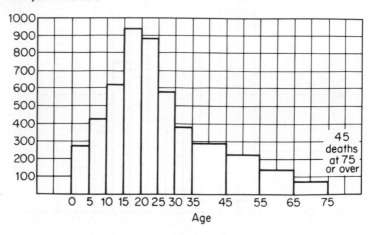

Fig. 3–5. Histogram for data of Table 3–5.

3.5. Frequency Polygons

The construction of a frequency polygon involves only elementary ideas, yet it is of considerable importance in obtaining a first notion of the nature of

Table 3–6

DISTRIBUTION OF DIVORCE CASES IN CHICAGO, 1919, BY FIVE-YEAR PERIODS OF MARRIED LIFE (Ernest R. Mowrer, *Family Disorganization.* Chicago University Press, 1927, p. 86)

Years of marriage	No. of divorces
0– 4	3164
5– 9	1515
10–14	743
15–19	348
20–24	191
25–29	70
30–34	25
35–39	15
40–44	4

a given frequency distribution. There are several important and widely-different types of distributions. Figures 3–1, 3, 4, and 5 are all of the same general type, roughly symmetrical with respect to a definite, somewhat central peak. They illustrate the most numerous and most important class of frequency distribution. Figure 3–6 (Table 3–6) belongs to the *J-type* of distribution, which is fairly common. (The J is here reversed.)

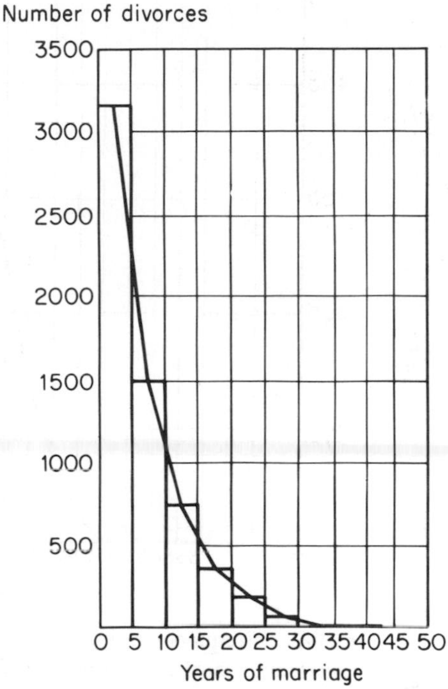

The relative proportions of a frequency polygon or of any graph are dependent upon the size of the horizontal unit compared with the vertical unit. Too small a horizontal unit makes the common form of frequency polygon seem narrow and tall; too large a unit makes it broad and flat. We might say for guidance that the rectangle enclosing the graph should ordinarily range from a height equal to about half the length to a height about equal to the length. A ratio of two to three is usually good. The graph itself should fairly well fill the rectangle.

In any graphical representation of statistical data, the diagram should contain the following items:

a. A label or title that completely describes the nature and source of the data. This should be printed so as to be read and understood easily.

b. The scale of measurement clearly indicated along the appropriate axes. Horizontal labels are preferable.

c. An accurately constructed figure of adequate size and proper proportions.

Fig. 3–6. Histogram and frequency polygon for data of Table 3–6.

3.6. Applications

It has been remarked that many frequency distributions yield frequency polygons that are characterized by a certain degree of symmetry with respect to a somewhat central peak. In particular, biological data involving measurements of length, width, weight, and so forth, are usually of this character. Sometimes this symmetry is very nearly perfect, as in the case of Figure 3–7. In other cases, the data may yield a *skew* polygon.

The class that corresponds to the maximum frequency in a distribution or to the peak of a polygon is called the *modal class*. For fairly symmetric

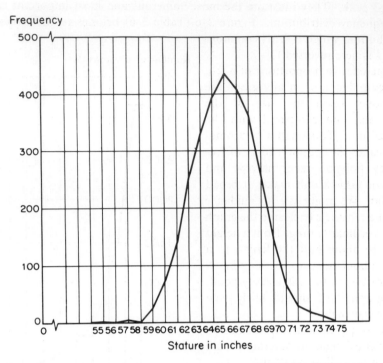

Fig. 3–7. Frequency polygon for the statures of 2984 English criminals. (Data from Charles Goring, *The English Convict*, H.M.S. Office, 1913, p. 386, slightly modified.)

distributions, it should be obvious that the average value will be somewhere within or near the modal class. From these facts it is often possible to derive important conclusions concerning the nature of observed data by reference to the frequency polygon alone. The following two illustrations bear on this point.

EXAMPLE 1. From *Genetics*, by Herbert E. Walter, 1919, p. 48 (by permission of The Macmillan Company, publishers):

Sometimes two conspicuous modes make their appearance in a frequency polygon, as Jennings found, for example, in measuring the body width of a population of the protozoan *Paramecium*. (Figure 3–8.)

It was subsequently found that the two modes in this polygon were due to the fact that the material in question was a mixture of two closely related species, *Paramecium aurelia* and *Paramecium caudatum*, the individuals of which arranged themselves around their own mean in each instance.

Although such an explanation does not always turn out to be the right one, the biometrician is led to suspect when a two or more moded polygon appears that he is dealing with a mixture of more than one kind of material, each of which fluctuates around its own average.

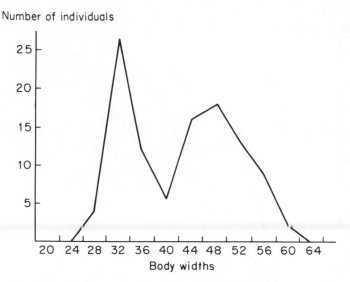

Fig. 3–8. Frequency polygons for body widths of paramecium.

EXAMPLE 2. From *Studies in Human Biology*, by Raymond Pearl, Williams and Wilkins, 1924, pp. 367–369.

Comparison of the polygons in Figure 3–9 would lead one to infer that the mothers of non-tuberculous individuals live, on an average, longer than do mothers of tuberculous individuals. This conclusion has been further justified by an analysis of the numerical data. The two peaks of the lower polygon suggest that the mothers themselves were either tuberculous (lower peak) with a tendency to die early, or non-tuberculous (upper peak) with a tendency to die normally.

3.7. Cumulative Frequency

It is common experience to hear questions of the following sort: How many tons of shipping did the United States lose during the first six months of World War II? How many students received grades lower than seventy per cent? What per cent of American girls twenty years of age weigh less than 110 pounds? Such questions are designed to elicit information of a special and significant type and usually require, for their answers, certain partial sums obtained from data classified according to some convenient

scheme. In statistics, these modes of expressing important facts center about
the concept of *cumulative frequency*.

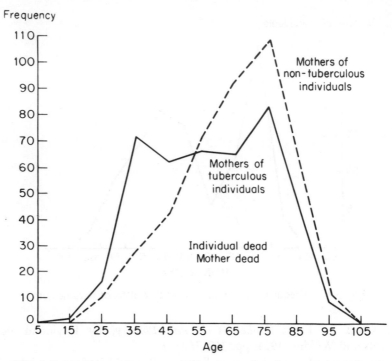

Fig. 3–9.   Frequency polygons showing the age at death of mothers of
tuberculous and non-tuberculous individuals (themselves dead).

The cumulative frequency (*abbreviated to* **cum f**) *of the* **m**th *class in a
frequency distribution is the sum of the frequencies, beginning with the first,*
f_1, *and ending with the* **m**th, f_m. *Briefly,*

$$cum\,f_m = \sum_{i=1}^{m} f_i.$$

From the familiar data of Table 3–3, we construct Table 3–7. The latter
includes the boundary values of each class as well as the cumulative frequencies. For example, cum $f_1 = 3$, cum $f_2 = f_1 + f_2 = 3 + 9 = 12$; cum $f_3 =
f_1 + f_2 + f_3 = 12 + 29 = 41$; and so on. Of course, cum $f_n = N$; so
that in Table 3–7, cum $f_{12} = 462$. When, for example, we say that
cum $f_5 = 221$ we mean that 221 head lengths were less than 191.5
millimeters.

Table 3–7

CUMULATIVE FREQUENCY TABLE OF HEAD LENGTHS

Mid-value x_i	Frequency f_i	Boundary	Cum f_i
		171.5	0
173.5	3		
		175.5	3
177.5	9		
		179.5	12
181.5	29		
		183.5	41
185.5	76		
		187.5	117
189.5	104		
		191.5	221
193.5	110		
		195.5	331
197.5	88		
		199.5	419
201.5	30		
		203.5	449
205.5	6		
		207.5	455
209.5	4		
		211.5	459
213.5	2		
		215.5	461
217.5	1		
		219.5	462
Total	462		

3.8. The Cumulative Frequency Diagram

The graph of the cumulative frequencies, sometimes called an *ogive*,[1] is obtained by plotting the cumulative frequencies against the boundary values or end-points of the class intervals. The plotted points are connected by line segments or by a smooth curve. From Figure 3–10, one can readily estimate the number of head lengths or the proportion or percentage of them less than any assigned value. See Section 4.7(3) for other uses. By "accumulating" the frequencies from the opposite end of the frequency table, we derive a "more than" table instead of a "less than" table, and obtain a descending graph instead of an ascending one. The form of the diagram is not appreciably affected by the presence of unequal class intervals, but, of course, care must be exerted to take account of them in marking off the horizontal scale.

[1] The shape of this curve is not usually that suggested by the corresponding architectural term, which denotes a diagonal rib or pointed arch. The statistical word *ogive* is derived more directly from the word *ogee* (O. G.), which designates an S-shaped molding.

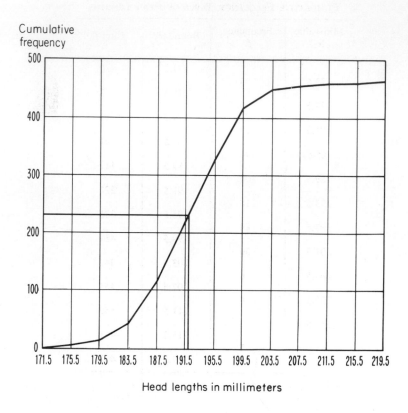

Fig. 3–10.   Ogive illustrating the graphical determination of the
median head length.

EXERCISES

1. The marks in Higher Mathematics received by candidates in a British Civil Service Examination are given in Table 3–8. Using class intervals 0–19, 20–39, and so on, construct a frequency table similar to that in Section 3.1.

Table 3–8

MARKS IN HIGHER MATHEMATICS RECEIVED BY CANDIDATES IN THE BRITISH CIVIL SERVICE EXAMINATION OF 1937 (MAXIMUM MARK, 200 POINTS)
(From *Questions, Papers, and Table of Results of the Competition Held in April 1937*, Civil Service Commission, Executive Group. By permission of the Controller of Her Britannic Majesty's Stationery Office)

168	153	115	105	73	78
185	140	128	90	95	90
150	143	85	125	93	60
135	133	128	123	13	23
168	125	100	85	90	20
155	155	133	75	53	48
118	130	100	130	70	33
165	170	120	130	90	48
153	145	120	108	90	25
140	138	115	85	98	20
143	133	138	53	88	80
160	140	73	98	83	18
145	108	123	73	40	30
130	130	150	93	75	33
125	95	135	88	25	28
135	143	48	85	68	63
160	90	123	113	43	20
115	118	70	113	78	50
138	135	103	65	65	33
140	163	90	100	50	3

2. Using class intervals of one inch, construct a frequency table like that in Section 3.1 for the heights in Table 3–9.

3. Do the preceding exercise for the weights. Use class intervals of 10 pounds.

4. (a) What are the class boundaries in Exercises 6–12?
   (b) What are the class marks (mid-values) in Exercises 5–12?
   (c) Select from the frequency distributions of Exercises 5–14 those that are discrete.

Table 3–9 (To accompany Exercises 2 and 3)

HEIGHTS IN INCHES AND WEIGHTS IN POUNDS OF 285 BOSTON
UNIVERSITY WOMEN STUDENTS

Hts.	Wts.	Hts.	Wts.	Hts.	Wts.	Hts.	Wts.	Hts.	Wts.	Hts.	Wts.	Hts.	Wts.
61	122	65	124	63	128	66	112	65	146	63	119	65	128
65	132	72	137	67	130	62	102	64	109	59	113	64	128
62	125	62	115	66	122	59	107	63	106	62	102	70	189
62	121	66	132	67	129	66	120	68	123	66	130	63	120
64	120	64	132	66	152	66	130	64	100	60	96	64	124
64	122	64	123	63	121	60	104	61	113	68	128	62	118
64	118	62	100	62	123	66	140	66	141	67	149	63	113
64	111	68	118	60	105	61	116	65	119	70	176	64	135
63	113	63	124	59	150	62	117	61	109	62	108	64	122
67	152	65	130	65	134	67	200	66	135	67	123	66	143
64	104	65	120	61	111	68	158	65	116	60	103	66	149
63	133	67	129	64	116	64	125	65	130	66	118	62	134
63	123	63	153	68	136	61	127	66	171	65	121	62	123
58	103	62	108	63	144	66	134	61	131	66	156	69	135
63	110	66	120	63	121	69	161	63	102	64	144	62	123
64	180	65	138	60	107	62	112	61	122	62	126	67	158
68	164	66	117	67	137	66	118	67	115	68	142	65	132
64	131	64	119	63	131	65	104	63	126	66	112	63	105
68	142	63	109	62	122	62	131	66	120	63	196	65	127
61	124	63	115	64	108	62	144	59	117	69	138	60	108
65	134	64	122	64	124	62	115	61	95	67	170	64	133
61	121	63	113	67	135	63	137	64	129	65	118	63	130
66	135	63	106	65	140	64	100	61	133	64	108	59	107
61	112	67	131	63	134	65	146	65	113	65	117	67	147
65	129	63	138	66	136	61	93	60	95	62	100	62	115
64	127	69	131	68	137	63	112	65	129	64	121	65	125
62	122	66	97	60	108	64	143	64	114	67	145	62	122
66	141	64	130	64	137	60	104	64	120	67	111	66	152
64	108	62	112	62	143	63	124	68	140	67	131	63	120
63	116	65	129	62	120	60	106	59	88	64	121	63	143
59	98	66	120	63	115	60	88	66	136	69	137	61	97
67	128	63	123	71	125	65	130	62	101	66	120	65	118
65	145	62	95	64	156	62	96	64	119	63	111	65	116
66	106	66	120	61	109	65	118	59	118	65	119	64	112
65	130	63	116	66	103	62	119	61	95	65	130	66	117
64	119	66	121	63	131	62	115	65	123	64	143	69	153
63	108	59	100	67	141	66	129	62	103	61	126	63	132
65	178	64	136	62	120	64	158	64	133	66	142	63	141
62	130	65	138	64	115	63	120	65	124	63	127	57	178
63	120	61	132	62	114	64	146	65	114	64	131		
63	120	64	125	64	116	66	128	61	124	67	148		

In Exercises 5–14 construct histograms (unless vertical line charts are more appropriate) and frequency polygons for the distributions.

5. The salaries of a group of employees:

Salaries	Number
$3000–3999	4
4000–4999	11
5000–5999	24
6000–6999	38
7000–7999	21
8000–8999	10
9000–9999	2

6. The brain-weights of adult Swedish males (from Raymond Pearl, *Studies in Human Biology*, Williams and Wilkins, Baltimore, 1924, p. 47):

Grams of brain-weight	Observed
Under 1100	0
1100–1149	1
1150–1199	10
1200–1249	21
1250–1299	44
1300–1349	53
1350–1399	86
1400–1449	72
1450–1499	60
1500–1549	28
1550–1599	25
1600–1649	12
1650–1699	3
1700–1749	1
1750 and over	0
Total	416

7. The age distribution (slightly modified) of admirals of the line of the United States Navy on active duty May 1, 1945. (See Section 3.2.)

Age	Number
40–44	1
45–49	37
50–54	75
55–59	87
60–64	44
65–69	22
70–74	6
75–79	1

8. Intelligence quotients of runaway boys (from Clairette P. Armstrong, *660 Runaway Boys*, Richard C. Badger, Gorham Press, Boston, 1932, p. 31):

I.Q.	Number
30– 39	2
40– 49	6
50– 59	60
60– 69	140
70– 79	184
80– 89	139
90– 99	78
100–109	37
110–119	14

9. Sizes of shoes worn by 235 college girls (original data):

Size	No. of girls
$2\frac{1}{2}$	2
3	2
$3\frac{1}{2}$	15
4	21
$4\frac{1}{2}$	21
5	36
$5\frac{1}{2}$	48
6	42
$6\frac{1}{2}$	26
7	15
$7\frac{1}{2}$	5
8	1
$8\frac{1}{2}$	1

10. Wages per day of male employees in 1860 (from Wesley C. Mitchell, *A History of the Greenbacks*, University of Chicago, 1903, p. 293):

Wage classes	Number
$0.25–0.49	118
0.50–0.74	123
0.75–0.99	599
1.00–1.24	2186
1.25–1.49	542
1.50–1.74	609
1.75–1.99	184
2.00–2.24	628
2.25–2.49	66
2.50–2.74	23
2.75–2.99	1
3.00–3.24	28
3.25–3.49	1
3.50+	3
	5111

11. The weights in milligrams of the dry contents of a certain type of ampul:

Weights	Number
80– 84	2
85– 89	7
90– 94	18
95– 99	32
100–104	14
105–109	10
110–114	3
115–119	1
120–124	1

12. Thicknesses in inches of steel washers:

Thickness	Frequency
0.095	1
0.096	4
0.097	9
0.098	15
0.099	20
0.100	24
0.101	21
0.102	16
0.103	7
0.104	3
0.105	2

13. Average weekly earnings during 1936 of urban Negro workers in the Mountain Region (*The Urban Negro Worker in the United States*, 1925–1936, Vol. 1, U.S. Department of Interior, p. 116):

Average weekly earnings	Total
Less than $5	38
$ 5–9	17
10–14	32
15–19	25
20–24	24
25–29	10
30–34	8
35–39	9
40–44	6
45–49	1
50–74	1
75–99	2
	173

14. A mathematics achievement test contained 40 questions for which the answers were marked either right or wrong. The distribution below summarizes the results.

No. answers correct	Frequency
0– 2	0
3– 5	1
6– 8	3
9–11	10
12–14	11
15–17	17
18–20	24
21–23	25
24–26	22
27–29	16
30–32	7
33–35	3
36–40	0
Total	139

15. Construct frequency polygons, as in Figure 3–9, for the distributions of weight, per mille (*a*) at mobilization, 1917–1918, and (*b*) at demobilization, 1919, of United States Army Troops. Compare the polygons and draw conclusions. (From C. B. Davenport and A. G. Love, *The Medical Department of the United States Army in the World War*, Vol. 15, *Statistics*, Part I, "Army Anthropology," 1921, p. 121.)

Class range	Number per mille	
	At mob.	At dem.
90– 99	0.21	
100–109	11.27	5.206
110–119	72.42	41.602
120–129	170.76	132.605
130–139	238.32	222.553
140–149	217.25	235.943
150–159	144.85	177.641
160–169	79.29	104.061
170–179	36.37	48.003
180–189	15.96	20.587
190–199	7.92	7.246
200 or over	5.40	4.561

16. The numbers of items answered correctly in the Beta examination. (Adapted from the "Report on the Mathematics Attainment Test of June 1936," by J. M. Stalnaker. College Entrance Examination Board, *Research Bulletin No. 7.*)

No. of items correct	Frequency
80–85	3
74–79	5
68–73	9
62–67	13
56–61	51
50–55	72
44–49	102
38–43	98
32–37	80
26–31	51
20–25	34
14–19	10
8–13	2
2– 7	1
	531

17. The Statistical Abstract of the United States for 1959 gives the following annual death rates for 1958.

Ages	Death rate per 1000
Under 1 yr.	29.8
1– 4	1.1
5–14	0.5
15–24	1.1
25–34	1.4
35–44	3.0
45–54	7.5
55–64	17.3
65–74	41.3
75–84	87.8
85 and over	202.7

Is there anything misleading or fallacious connected with Figure 3–11, constructed from these data?

18. Write the limits for the first two and the last two class intervals you would select in constructing a frequency table for each of the following:
(*a*) The lives of 200 electric light bulbs. Shortest life, 63 hours; longest life, 1281 hours.
(*b*) The weights of 160 aspirin tablets. Smallest, 4.82 grains; largest, 5.13 grains.
(*c*) The daily piece-work output of 135 workers. Smallest, 59; largest, 95.
(*d*) Sizes of 4000 pairs of men's socks. Smallest, 6; largest, 13.

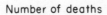

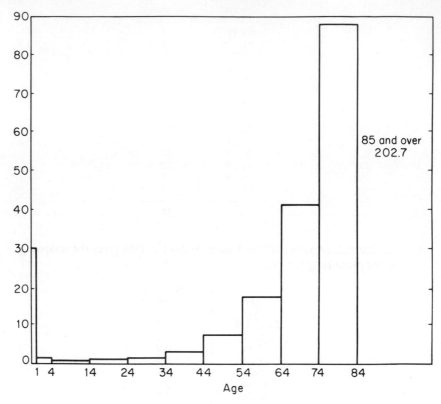

Fig. 3–11.   Frequency polygon for number of deaths by age groups.

19. Construct cumulative frequency diagrams for the following distributions:

(*a*) Salaries of employees (Ex. 5).

(*b*) Brain-weights of Swedish males (Ex. 6).

(*c*) Ages of admirals (Ex. 7).

(*d*) I.Q.'s of runaway boys (Ex. 8).

(*e*) Weights of ampuls (Ex. 11).

(*f*) Thicknesses of washers (Ex. 12).

(*g*) Numbers of correct answers (Ex. 14).

20. What conclusions can you draw from Figure 3–12?

Cumulative per cent

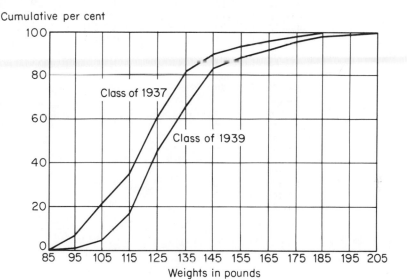

Fig. 3–12. Cumulative percentage distributions of weight in pounds of two classes of college women students.

21. A visiting nurse association asked its nurses to record the number of minutes spent on each call. As a result 1576 calls were recorded during a stated period. It was expected that these data, when compiled into a frequency table and then represented by a histogram, would be of the general bell-shaped form with possibly some skewness present. Table A below was constructed and exhibited great irregularity in the pattern of distribution. Then Table B, with doubled class widths, was made, and although an improvement in pattern was noted, it was still unsatisfactory. (a) Why was not the expectation of the bell-form realized? (b) Assuming that you had at hand the original data, how would you have constructed this table?

Table 3–12

A			B	
No. minutes	No. visits		No. minutes	No. visits
0– 4	52		0– 8	121
5– 8	69		9– 16	236
9– 12	92		17– 24	189
13– 16	144		25– 32	301
17– 20	154		33– 40	268
21– 24	35		41– 48	130
25– 28	97		49– 56	143
29– 32	204		57– 64	90
33– 36	133		65– 72	57
37– 40	135		73– 80	21
41– 44	17		81– 88	9
45– 48	113		89– 96	6
49– 52	90		97–104	3
53– 56	53		105–112	2
57– 60	83			
61– 64	7			
65– 68	37			
69– 72	20			
73– 76	15			
77– 80	6			
81– 84	3			
85– 88	6			
89– 92	4			
93– 96	2			
97–100	3			
101–104	0			
105–108	0			
109–112	2			

THE STATISTICS OF
FREQUENCY DISTRIBUTIONS

4

"Seeing there is nothing (right well beloved students in the Mathematickes) that is so troublesome to Mathematicall practice, nor that doth more molest and hinder Calculators than the Multiplications, Divisions, square and cubical Extractions of great numbers, which, besides the tedious expense of time, are for the most part subject to many slippery errors... and having thought upon many things to this purpose, I found at length some excellent briefe rules."

JOHN NAPIER
Mirifici Logarithmorum Canonis Descriptio (1614)

4.1. The Arithmetic Mean

The arithmetic mean of a group of numbers arranged in a frequency distribution is conveniently conceived as a weighted mean, (Section 2.6) where the variates are the mid-values of the intervals and the weights are the corresponding frequencies. Formula (2.3) becomes, then,

$$\bar{x} = \frac{1}{N} \sum_{i=1}^{n} f_i x_i \qquad (4.1)$$

where f_i is the frequency of the class whose mid-value is x_i, n is the number of classes, and $N = \sum_{i=1}^{n} f_i$, the total frequency. Formulas (2.1) and (4.1), above, represent essentially the same thing, for the latter merely takes into account the fact that certain variates occur repeatedly, the number of repetitions being designated as the frequency f.

Formula (4.1) gives precisely the same value for $\bar{x}$ as Formula (2.1), provided the arithmetic mean for each class interval is exactly the mid-value of the interval. This, however, is rarely the case, save possibly in a discrete set. Nevertheless, most frequency distributions in practice are of the roughly symmetrical type, where the larger frequencies build up toward a maximum somewhere near the middle of the range of values. Thus, the mean value of each class interval in the lower half of the range is likely to be above the mid-value of the interval, and the mean value of each class interval in the upper half of the range is likely to be below the mid-value of the interval. As a consequence of these two compensating effects, Formula (4.1) will, in general, be accurate enough for most purposes, especially when N is large and the class interval reasonably small. In the case of the J-type of distribution, there will be a distinct bias given to the mean in the direction of the smallest class frequencies, and unless the class intervals are quite small, the error committed may be serious. For other more or less irregular distributions, no general rule can be stated save that small class intervals will increase the accuracy of Formula (4.1).

The calculation of the mean from a frequency table is expedited by *coding* the mid-values, x_i. An *assumed* or *provisional* mean and a convenient divisor are used. The assumed mean is a mid-value taken somewhere near the value that the true mean is estimated to have and the divisor is chosen equal to the width of the class interval. The process of transforming variates to values more amenable to calculation is called *coding*, and the transformed variates are called *coded values*.

We designate the provisional mean by x_0, the divisor by k, and let the coded value be

$$u_i = \frac{x_i - x_0}{k}. \tag{4.2}$$

We take as an example the data of Table 4–1. We select the somewhat central mid-value 193.5 as the provisional mean, x_0, from which the coded values or *unit deviations*, u_i, are recorded in column 3. These deviations are always integers, positive or negative (or zero), since the class intervals are reduced to unit width. For example,

$$u_1 = \frac{x_1 - x_0}{k} = \frac{173.5 - 193.5}{4} = -5$$

and

$$u_9 = \frac{x_9 - x_0}{k} = \frac{205.5 - 193.5}{4} = 3.$$

Instead of computing Σfx, as required by Formula (4.1), we compute Σfu, as shown in column 4, by multiplying the corresponding values of f and u in columns 2 and 3 and then dividing by the total frequency, 462. The result is the arithmetic mean of the u's, so that

$$\bar{u} = \frac{1}{N} \Sigma fu$$

$$= \frac{-196}{462} = -0.424.$$

We may write (4.2) as,

$$ku_i = x_i - x_0$$

whence,

$$x_i = ku_i + x_0. \tag{4.3}$$

Multiplying by f_i and then taking $1/N$th the sum of both members of (4.3) we have, by making use of the theorems of Section 2.2,

$$\frac{1}{N} \Sigma f_i x_i = \frac{1}{N} \Sigma f_i(ku_i + x_0)$$

$$= \frac{1}{N} \Sigma k f_i u_i + \frac{1}{N} \Sigma f_i x_0$$

$$= k \cdot \frac{1}{N} \Sigma f_i u_i + \frac{1}{N} \cdot N x_0;$$

whence,

$$\bar{x} = k\bar{u} + x_0. \tag{4.4}$$

Applying this formula to the head-length data we have

$$\bar{x} = 4(-0.424) + 193.5$$

$$- 191.8.$$

To illustrate further the method of coding, let us refer to Table 3–2. The class boundaries for the monthly bills for electricity are 2.995–3.495, 3.495–3.995, and so on. We select the mid-value, 6.245, of the class interval, 5.995–6.495, as the provisional mean x_0, corresponding to $u_0 = 0$. The u's then range from -6 at the top of the column to 5 at the bottom. If the calculations are carried out as in Table 4–1, we obtain $\Sigma fu = 8$. Then $\bar{u} = \frac{8}{66} = 0.121$. By Formula (4.4) $\bar{x} = (0.50 \times 0.121) + 6.245 = 6.3055$. The mean monthly bill is, therefore, $6.31. It is interesting to note that if the 66 individual amounts in Table 3–1 were added and the sum divided by 66, we would obtain $6.33 as the mean. The error committed by using a frequency distribution to calculate the mean is generally negligibly small.

4.2. Remarks Concerning the Computation

(1) *Checking.* The computation of the arithmetic mean may be verified by selecting a different provisional mean, x_0, and then applying the method of the preceding section. Referring to columns 6 and 7 of Table 4–1, we have:

$$\bar{u} = \frac{266}{462} = 0.576 \quad \text{and} \quad x_0 = 189.5,$$

so that
$$\bar{x} = k\bar{u} + x_0$$

$$= 4(0.576) + 189.5 = 191.8.$$

This agrees with the result previously obtained.

If the new provisional mean selected is one just above the old the following relation should hold:

$$\boldsymbol{\Sigma f(u + 1) = \Sigma fu + \Sigma f}$$

$$\boldsymbol{= \Sigma fu + N.}$$

Thus, in Table 4–1,

$$\Sigma f(u + 1) = 266; \quad \Sigma fu = -196; \quad N = 462.$$

hence,

$$266 = -196 + 462.$$

Table 4–1
COMPUTATION OF THE MEAN AND STANDARD DEVIATION OF HEAD LENGTHS

(1)	(2)	(3)	(4)	(5)	(6)	(7)	(8)
Mid-value						Check	
x	f	u	fu	fu^2	$u + 1$	$f(u + 1)$	$f(u + 1)^2$
173.5	3	−5	−15	75	−4	−12	48
177.5	9	−4	−36	144	−3	−27	81
181.5	29	−3	−87	261	−2	−58	116
185.5	76	−2	−152	304	−1	−76	76
189.5	104	−1	−104	104	0	0	0
193.5	110	0	0	0	1	110	110
197.5	88	1	88	88	2	176	352
201.5	30	2	60	120	3	90	270
205.5	6	3	18	54	4	24	96
209.5	4	4	16	64	5	20	100
213.5	2	5	10	50	6	12	72
217.5	1	6	6	36	7	7	49
			−394			−173	
			198			439	
Totals	462		−196	1300		266	1370

(2) *Significant digits.* The head lengths of the 462 English criminals were measured to the nearest millimeter, and hence with an accuracy of three significant digits. Their mean value has been computed to the nearest tenth of a millimeter and found to be 191.8. According to the remark in the last paragraph of Section 2.3, we have good justification for saving four digits rather than three, since the total frequency, 462, is fairly large. However, the question of the precision of the arithmetic mean is not one that can be answered satisfactorily at this point. (See Sections 8.2 and 10.5.) A good working rule is to save one more digit than occurs in most of the individual items from which the frequency table is constructed.

(3) *Unequal class intervals.* In some frequency distributions, the class intervals are not all of the same width, so that the method of the preceding section must be either modified or abandoned. If there is much irregularity among the widths, it is better not to attempt to make use of *unit* deviations, but to use *non-unit* deviations from a convenient provisional mean. This means, merely, that k is omitted in Formula (4.2).

If there are but two different class widths, a desirable method is to select as the provisional mean, x_0, a suitable number lying between the two mid-values at which the change in width occurs.

Sometimes the frequency distribution may be divided into several parts in each of which the class interval is uniform, the mean computed for each part, and then the *weighted arithmetic* mean of these means taken. (See Section 2.6.)

(4) *Non-symmetrical distributions.* In computing the arithmetic mean from a frequency distribution, we assumed that all the values lying within a given class interval could be replaced by the mid-value. This is a valid assumption provided that the distribution is a fairly symmetric one; but if the distribution exhibits considerable skewness, or if the data follow a highly irregular pattern, the methods of computation just outlined may be no longer valid. It may be necessary to employ very small class intervals or to dispense with the distribution and to revert to the raw data from which it was constructed. Occasionally more advanced techniques may be employed. Frequently the arithmetic mean must be abandoned in favor of a different, more representative average.

(5) When an electric desk calculator or other high-speed computer is available, one may often dispense with the coding of observations.

4.3. The Variance and Standard Deviation

The computation of s^2 and s from a frequency distribution follows the same general pattern as that for $\bar{x}$ and may be regarded as a logical extension of it. The variance of the coded values, s_u^2, is obtained from Formula (2.7) by introducing the weighting factors, the class frequencies, as in the case of $\bar{x}$.

Then,

$$s_u^2 = \frac{1}{N} \Sigma f u^2 - \bar{u}^2. \tag{4.6}$$

If we subtract Equation (4.4) from (4.3) we have

$$x_i - \bar{x} = k(u_i - \bar{u}). \tag{4.7}$$

Squaring, multiplying by f_i, and taking $1/N$th the sum of each member we obtain,

$$\frac{1}{N} \Sigma f_i(x_i - \bar{x})^2 = \frac{1}{N} \Sigma k^2 f_i(u_i - \bar{u})^2$$

$$= k^2 \cdot \frac{1}{N} \Sigma f_i(u_i - \bar{u})^2$$

or $$s_x^2 = k^2 s_u^2. \tag{4.8}$$

Also $$s_x = k s_u, \text{ where } k > 0. \tag{4.9}$$

It is apparent that the use of a provisional mean, x_0, does not affect the variance, but that the latter *is* affected by a constant divisor or multiplier of the variates. Thus, the variance of x is the same as the variance of $x - x_0$, but the variance of x is k^2 times the variance of x/k.

As an illustration of the use of Formulas (4.6) and (4.9) let us calculate the standard deviation of the head lengths (Table 4–1):

$$s_u^2 = \frac{1300}{462} - (-0.424)^2 = 2.63;$$

$$s_u = \sqrt{2.63} = 1.62.$$

From (4.9) $$s_x = 4(1.62) = 6.48.$$

For the data of Table 3–2, with $x_0 = 6.245$ as in Section 4.1, we find $\Sigma f u^2 = 292$. Then

$$s_u^2 = {}^{292}\!/_{66} - (0.121)^2 = 4.41.$$

$$s_u = \sqrt{4.41} = 2.10,$$

hence by (4.9)

$$s_x = 0.50 \times 2.10 = 1.05.$$

4.4. Remarks Concerning the Computation of s^2 and s

(1) *Checking.* As in the case of the arithmetic mean, the value of the standard deviation found from a frequency distribution may be verified by

selecting a new provisional mean adjacent to the original one chosen. Since

$$\Sigma f(u + 1)^2 = \Sigma f(u^2 + 2u + 1) = \Sigma fu^2 + 2\Sigma fu + \Sigma f, \quad (4.10)$$

we may compare, in Table 4–1, the total in column 8 with the total of column 5 added to twice the total of column 4 plus the total of column 2,

$$1370 = 1300 + 2(-196) + 462 = 1370.$$

(2) *Significant digits.* In general, an accuracy of three significant digits will be satisfactory for most practical problems connected with the standard deviation. For a description of the variability of head lengths, the value $s = 6.5$ mm would be generally sufficient and justifiable.

(3) *Unequal class intervals.* If a frequency distribution contains groups of unequal class intervals, the variance may often be calculated advantageously by making a wise choice of x_0 and by taking proper account of the inequalities in the intervals. This method works well when only two differing groups exist.

In other cases, particularly when there are more than two groups with unequal class intervals, it may be wiser to compute the mean and variance of each group separately and then to apply the following formula:

$$Ns^2 = \sum_{i=1}^{m} N_i s_i^2 + \sum_{i=1}^{m} N_i(\bar{x}_i - \bar{x})^2 \quad (4.11)$$

where the ith subgroup has a total frequency, N_i, a mean, $\bar{x}_i$, and a variance, s_i^2. $N = \sum_{i=1}^{m} N_i$.

4.5. Uses of the Standard Deviation

There is no question that s is the most widely used measure of dispersion. A relatively small value of s denotes close clustering about the mean; a relatively large value, wide scattering about the mean.

4.6. The Median

The median may be readily computed from a frequency table with the aid of a column of cumulative frequencies. The definition that follows may be shown to be a natural extension of the definition in Section 2–7. (See Chapter IV of F. Zizek, *Statistical Averages*, New York: Henry Holt, 1913.)

The median of a frequency distribution is the value of the variable, x, that corresponds to the cumulative frequency, ½ N.

This value may be estimated by the method of interpolation by proportional parts. In Table 3–7, $½ N = 231$, cum $f_5 = 221$, and $231 - 221 = 10$;

hence, the value of x corresponding to cum $f = 231$ is the tenth value among the 110 head lengths assumed to range in order from 191.5 to 195.5 millimeters. Hence, we take $^{10}\!/_{110}$ of the class interval, 4, and add the result to 191.5, the boundary corresponding to cum $f_5 = 221$. Thus, the median equals:

$$(^{10}\!/_{110} \times 4) + 191.5 = 191.9 \text{ (millimeters)}.$$

This means that there were as many head lengths greater than 191.9 millimeters as there were less.

As an additional example of the method of interpolating for the median, we use again the data of Table 3–2. $\frac{1}{2} N = 33$; cum $f_6 = 21$; cum $f_7 = 39$; $f_7 = 18$. The value of the bill corresponding to cum $f = 33$ is the twelfth among the 18 amounts assumed to range in order from 5.995 to 6.495. Then $(^{12}\!/_{18} \times 0.50) + 5.995 = 6.328$. Thus the median bill amounts to $6.33.

4.7. Remarks Concerning the Computation of the Median

(1) *Significant digits.* The statements concerning the precision of the mean (Section 4.2) are, in general, applicable also to the median. For a relatively small number of items, say 40 or less, the number of digits to be used in writing the median value is usually equal to that occurring in most of the individual items. For a relatively large number, say between 40 and 400, an extra digit may be saved with some degree of justification.

(2) *Distribution within an interval.* The method of interpolation just described assumes that the variates belonging to a class are uniformly spaced within the class interval. Such a distribution is, of course, not strictly in accord with experience. The assumption is open to criticism, particularly if the total frequency is fairly small, if the class interval is wide, or if the distribution is not reasonably symmetric. Nevertheless, for most frequency distributions arising in practice, the method outlined is considered satisfactory.

(3) *Graphical computation.* A graphical method of interpolation may also be used. From the cumulative frequency polygon, the abscissa of the point corresponding to the ordinate, $\frac{1}{2} N$, is estimated. If more accurate results are desired, the broken line may be replaced by a smooth curve, From Figure 3–10, the median head length is seen to be equal to about 191.9, in this case the same as the value obtained by arithmetic interpolation.

(4) *Unequal class intervals.* The method of determination of the median of a frequency distribution by either arithmetic or graphic interpolation is still applicable when the class intervals are not uniform.

(5) *Discrete distributions.* In most frequency distributions of discrete variables, the method of interpolation is based upon the use of artificial class boundaries. For this reason and for other reasons, the median is usually

valueless in such distributions. The data of Table 4–2 form an example of a discontinuous distribution for which the median value, 6, has practically no statistical value.

Table 4–2

NUMBER OF MICE DYING PER GROUP OF SIX
UNDER CERTAIN SERUM TESTS

No. of mice dying	f_i	Cum f_i
0	42	42
1	31	73
2	13	86
3	7	93
4	9	102
5	71	173
6	355	528
	528	

4.8. Properties of the Median

The median of a frequency distribution varies somewhat with the method of grouping, that is, with the choice of class limits and widths of class intervals. This is an undesirable characteristic, but it offers no serious difficulties in practice.

The median is not readily affected by a few abnormal values, extremely small or extremely large, and this stability makes it a useful average in many cases. For example, in Table 3–2, the replacement of the last three electric current bills by three abnormally large ones, say each of $15, would in no wise affect the median, but would raise the arithmetic mean by an appreciable amount (about 29 cents.)

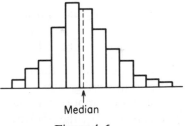

Median

Figure 4–I

A vertical line passing through the median of a frequency distribution divides its histogram into two parts of equal area. (See Figure 4–1.) This follows from the definition of the median and from the fact that the areas of the constituent rectangles are proportional to the corresponding class frequencies.

4.9. The Usefulness of the Median

Certain distributions quite definitely tend to be skew because of some natural limit at one end of the range of variation. Thus, the hourly wages of a certain class of workers might exhibit skewness due to a minimum wage law and the fact that many workers (theoretically at least) can attain exceedingly high wages. In such cases, the median might well be more truly typical of the average wage than the arithmetic mean.

The median is often useful when an accurate determination of the arithmetic mean is impossible. This occurs when a frequency table is "open" at one end or both ends. (See, for example, Exercise 3–10.) Indefinite classes at the ends of a distribution will not affect the position of the median.

The median is employed as a useful average throughout a wide range of statistical investigations. Among them the following may be mentioned: in mortality studies for finding the average length of life (probable lifetime), in various kinds of price and wage statistics, in toxicology for measuring the potency of a drug (see Exercise 12–9), in industrial quality control, in problems dealing with data that are arranged according to some qualitative criterion, and in nonparametric inference (Chapter 15).

4.10. The Percentiles

The mth percentile, P_m, of a frequency distribution is the value of the variable, x, that corresponds to the cumulative frequency, m per cent of N. It is conceived as the value for which m per cent of the variates are smaller and $(100 - m)$ per cent are larger. A percentile is computed by the familiar method of interpolation. Referring to Table 3–7, we compute the 30th percentile, P_{30}.

$$0.30N = 0.30 \times 462 = 138.6$$

$$P_{30} = \left(\frac{138.6 - 117}{104} \times 4\right) + 187.5 = 188.3.$$

Thus, 30 per cent of the head lengths were less than 188.3 millimeters and 70 per cent were greater.

Of major importance among the percentiles are the 25th percentile or *first quartile*, Q_1, the 75th percentile or *third qaurtile*, Q_3, and the *deciles*. The rth decile, D_r, corresponds to the $10r$-th percentile, where $r = 1, 2, 3, ..., 9$. Thus, the third decile, D_3, is the 30th percentile. The deciles divide the ordered set of variates into ten portions of equal frequency. They are employed occasionally to give in greater detail the range of distribution of the variates and to characterize the nature of a set of qualitatively ordered variates. An example of the latter might be a group of students arranged according to native ability.

For the head length data

$$\tfrac{1}{4} N = \tfrac{1}{4} (462) = 115.5;$$

$$Q_1 = \left(\frac{115.5 - 41}{76} \times 4\right) + 183.5 = 187.4.$$

One-fourth of all the 462 head lengths, then, were less than 187.4 millimeters.

Similarly,

$$\frac{3}{4} N = \frac{3}{4}(462) = 346.5;$$

$$Q_3 = \left(\frac{346.5 - 331}{88} \times 4\right) + 195.5 = 196.2.$$

All the remarks of Section 7 concerning the median hold without exception for the other percentiles. In particular, attention should be called to the graphical methods of estimating the percentiles.

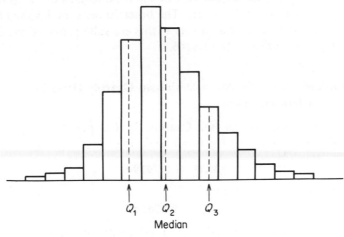

Figure 4–2

The vertical lines through the quartiles divide the area of the histogram into four equal areas (Figure 4–2). This means that the area to the left of Q_2, the area to the right of Q_2, the area between Q_1 and Q_3, and the sum of the two areas "outside" of Q_1 and Q_3, are each equal to 50 per cent of the total area.

The quartiles have important uses in connection with the measurement of the degree of dispersion of the data about the center and with the measurement of the skewness of the distribution.

As defined, a *quartile* is a point or a mark in an ordered set of values, but it is often used also to designate an interval or a range of values. We find it convenient to state, for example, that a value lies in the second quartile, that is, between the marks Q_1 and Q_2.

4.11. The Quartile Deviation

It has just been pointed out that 50 per cent of the total distribution is comprised of variates lying between the first and third quartiles. This suggests that one-half the interval $Q_3 - Q_1$ would measure the "average" deviation of the variates from the median or other intermediate value. We therefore define the *quartile deviation* or *semi-interquartile range*, *Q*, as:

$$Q = \tfrac{1}{2}(Q_3 - Q_1). \tag{4.12}$$

It is an appropriate measure of dispersion to accompany the median. Referring to Section 10, the quartile deviation for the head lengths of 462 criminals is

$$\tfrac{1}{2}(196.2 - 187.4) = 4.4.$$

This means, roughly, that 50 per cent of the head lengths deviated from the median by less than 4.4 millimeters. The other 50 per cent deviated by more than 4.4 millimeters. Such statements as this are valid provided the distribution is fairly symmetrical. (See Ex. 46.)

4.12. Relations Among the Mean Deviation, Quartile Deviation, and Standard Deviation

For normal distributions (see Chapter 7) it is possible to prove easily with the aid of calculus that the mean deviation

$$M.D. = 0.7979\sigma, \tag{4.13}$$

and that the quartile deviation

$$Q = 0.6745\sigma. \tag{4.14}$$

Thus, for fairly large samples that are moderately normal in form, we may expect the following approximate relations to hold:

$$M.D. = \tfrac{4}{5}\,s; \tag{4.15}$$

$$Q = \tfrac{2}{3}\,s. \tag{4.16}$$

For the data of head lengths, $M.D. = 5.2$, $Q = 4.4$, and $s = 6.5$. Then

$$\frac{M.D.}{s} = \frac{5.2}{6.5} = 0.80;$$

$$\frac{Q}{s} = \frac{4.4}{6.5} = 0.68.$$

In practice, Formulas (4.15) and (4.16) will be found to hold surprisingly well.

4.13. The Coefficient of Variation

The value of the standard deviation of a set of positive variates, grouped or ungrouped, depends, among other things, upon the units in which the variates are expressed. It is often desirable to compare the dispersion of two sets of positive variates measured in the same or in different units. Examples of data subject to such comparison might be the hourly wages of coal miners and of steel workers, or the lengths and weights of some living body. In order to reduce a statistical measure to a number that is independent of the unit of measurement used, this measure is divided by some other one expressed in the same unit. The result is a so-called *pure number* or *absolute measure*.

A common absolute measure of dispersion is the *coefficient of variation, V*, defined as:

$$V = \frac{s_x}{\bar{x}}, \tag{4.17}$$

or as:

$$V = 100 \frac{s_x}{\bar{x}} \text{ (per cent)}. \tag{4.18}$$

Thus, for the head lengths of criminals,

$$V = \frac{6.48}{191.8} = 0.0337.$$

This means that the standard deviation is about 3.37 per cent of the arithmetic mean.

The amount of dispersion of data seems to be connected, in a vague way, with the distance from the origin or zero point to the mean value; therefore, the coefficient of variation should be used only in cases where an origin is inherent in the data. Measurements of lengths, weights, hourly wages, prices of goods, school grades in per cent, and so forth, cannot be less than zero; hence, for such data, the quantity, V, has significance. On the other hand, temperature readings are entirely relative to the particular scale used. A reading of $0°$ Centigrade is equivalent to $32°$ Fahrenheit. The reason for the restriction of the use of V to positive variates should, by now, be apparent. Surely the presence of $\bar{x} = 0$ in Formula (4.17) or (4.18) ought to be embarrassing to the would-be statistician.

4.14. The Mode

The average discussed in this section was illustrated briefly in Section 2.1. It is generally most useful when the variates are numerous enough to make a frequency distribution possible. The mode marks a point corresponding to the highest frequency.

The modal class of a frequency distribution is the class corresponding to the maximum frequency. More generally (in order to cover the possibility of *multimodal* data, that is, of frequency distributions whose polygons have more than one distinct "peak") we may define the modal class as that class of greater frequency than any near-by class. In the data of Table 4–1, the modal class is the class whose mid-value is 193.5 mm. This fact is clearly shown graphically in Figure 3–1, where the tallest rectangle corresponds to the class interval 191.5 — 195.5.

The preceding definitions are usually useful, however, only in the case of discrete data. For example, if a certain shoe store sold more men's shoes of size 8 than any other size, 8 would be the modal size.

These definitions, save in cases of the type just cited, are open to serious criticism. In the first place, data may be assembled into class intervals in various ways, and the modal value will vary with the choice of the class intervals. In the second place, the above definitions concern a *modal class*, an interval, and not a *mode*, a single value. In Chapter 7 we shall introduce the concept of a *frequency curve*, the graph of the population from which the sample frequency distribution may be imagined to originate. Since we conceive of a frequency distribution as determining a frequency curve, the value of the variable, x, corresponding to the maximum point of this curve is precisely the mode of the distribution, and the maximum point on the curve rarely coincides exactly with the peak of the polygon. The accurate determination of the mode may require a knowledge of calculus.

Statistically, the mode represents the most typical, the most frequent, value. It is often confused popularly with the arithmetic mean.

4.15. Further Remarks on the Appropriateness of an Average

At the end of Chapter 2 the advantages and disadvantages of some of the measures of central tendency were discussed. The statements made there may be amplified with reference to frequency distributions.

For fairly symmetric distributions, the three averages are very nearly equal and may be considered to be equally valid. The selection, then, might be made solely on the basis of ease of computation or the degree of familiarity on the part of those for whom the average is intended. The arithmetic mean is easily understood and computed without difficulty, as a rule. Skew distributions of the type illustrated in Figure 4–4 do not always permit the mean to be as truly representative as it ordinarily is. If a fairly high degree of skewness is present, the mode or the median may be a better average. Thus, when the median differs markedly from the mean, the former is more likely to be a better average. The median, as has been pointed out, is useful when the frequency table is incomplete at an end, when a few extremely large or small variates are present, or when the classification is according to a qualitative order. In educational statistics, the desirability of dividing a class of pupils according to ability into two or more parts makes the median particularly useful at times. We may wish to know if Susie Smith stands in the upper half of the class or the lower. The median marks this division.

The mode is likely to have genuine practical value for discrete data. It is also singularly appropriate when one value or class interval is predominant.

4.16. Standard Deviates

Statistics involves, among other things, a comparative study of data. This study necessitates a reduction of different units of measurement to a common measure, or, better still, to pure numbers that are independent of the units used. For this and for other reasons, it has become universal practice to measure variates in terms of their standard deviation as a unit. We therefore introduce z, defined by the formula:

$$z = \frac{x - \bar{x}}{s_x}. \tag{4.19}$$

Thus z is a *standard deviate*, since it measures the deviation from the mean in terms of the standard deviation as a unit. For example, since for the head lengths of criminals $\bar{x} = 191.8$ and $s = 6.48$, a head length of 204.3 corresponds to a standard deviate,

$$z = \frac{204.3 - 191.8}{6.48} = 1.93.$$

Thus 204.3 exceeds the mean by 1.93 standard deviations.

4.17. Moments

Frequency distributions have already been described in two different ways. *Geometrically*, the common bell-shaped type has been characterized by its symmetry or skewness, by the location of certain important points such as the mean, median, and mode, and by its degree of dispersion. Such a geometric description is readily comprehended, but it is essentially qualitative and often fails to reveal significant latent properties. *Algebraically*, the attributes of a frequency distribution have been precisely measured with the aid of certain functions of the variates, among which the mean, $\frac{1}{N} \Sigma fx$, and the variance, $\frac{1}{N} \Sigma f(x - \bar{x})^2$, are of chief importance. A value, such as $\bar{x}$ or s, calculated from a given set of variates, is called a *statistic*. There are two other *statistics* that serve admirably to characterize frequency distributions. They are called the *third* and *fourth moments* of the distribution. Moments are not entirely unfamiliar to us, for the arithmetic mean is a first moment, and the variance is a second moment.

4.18. The Third Moment; Skewness

We define the *third standard moment*, a_3, by means of the formula

$$a_3 = \frac{1}{N} \sum f \left(\frac{x - \bar{x}}{s} \right)^3$$

$$= \frac{1}{N} \sum f z^3. \tag{4.20}$$

The choice of a_3 as a suitable measure of skewness depends upon the fact that in a perfectly symmetric distribution the cubed positive deviations (with

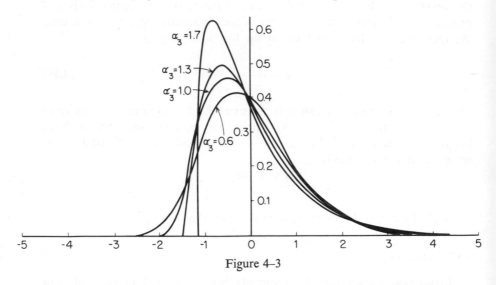

Figure 4–3

respect to the mean) are exactly balanced by the cubed negative deviations; thus, $a_3 = 0$. It follows that any non-zero third moment indicates some degree of asymmetry. If a_3 is positive, there must be more dispersion above the mean than below; if a_3 is negative, the reverse must be the case. Similar remarks apply to odd moments of higher order.

As an illustration, it may be shown that for the head lengths of criminals $a_3 = 0.176$.

The skewness of a theoretical distribution or population is symbolized by α_3 (read "alpha sub-3"). In asymmetric distributions, known as Pearson Type III distributions, α_3 ranges from -2 to $+2$, hence sample frequency distributions assumed to be drawn from such populations are expected to range similarly. Figure 4–3 shows several such curves and gives an idea of the degree of skewness associated with a given value of α_3. Figure 4–4 shows a skew frequency curve arising from extensive data gathered by the College Entrance Examination Board.

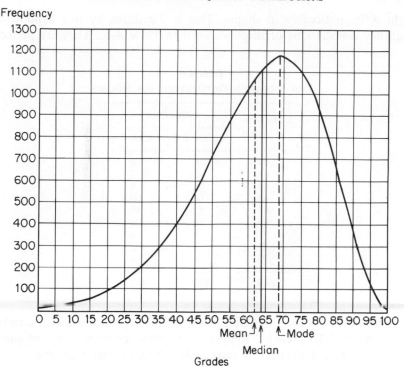

Fig. 4–4. Skew frequency curve of the ratings of approximately 10,000 candidates. (*Thirty-First Annual Report of the Secretary of the College Entrance Examination Board*, 1930, page 17. Published by the Board, New York, 1931. The curve shown in this report has been slightly modified by the addition of the median line.)

4.19. The Fourth Moment

The *fourth standard moment*, a_4, is defined by the formula

$$a_4 = \frac{1}{N} \sum f \left(\frac{x - \bar{x}}{s} \right)^4$$

$$= \frac{1}{N} \sum f z^4. \tag{4.21}$$

For the distribution of head lengths, we find that the fourth moment $a_4 = 3.67$.

It might appear that a sharply peaked polygon implies a relatively small value for the standard deviation, while a flat polygon implies a relatively large value; but these implications are by no means necessary. Two perfectly symmetric distributions with the same value for the standard deviation might differ noticeably in shape. This is illustrated by two hypothetical frequency distributions of heights shown in Table 4–3. They are perfectly

Table 4–3
TWO FICTITIOUS DISTRIBUTIONS FOR HEIGHTS

Heights in inches	Frequencies	
	Case 1	Case 2
59	1	0
61	2	2
63	5	8
65	16	20
67	52	40
69	16	20
71	5	8
73	2	2
75	1	0
Total	100	100

symmetric distributions, and each has a mean of 67 inches and a standard deviation of 2.37 inches. Reference to their corresponding frequency polygons (Figure 4–5) shows a marked difference in their peakedness.

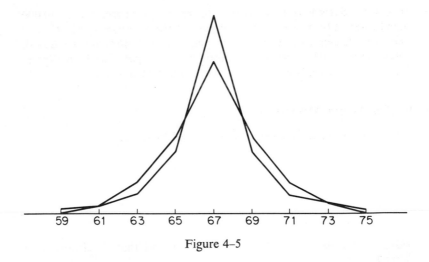

Figure 4–5

It is instructive to compute the fourth moments for these two frequency distributions. In Case 1, $a_4 = 5.24$, in Case 2, $a_4 = 3.16$. In a perfectly normal distribution $\alpha_4 = 3$, precisely. The moments a_3 and a_4 are useful in testing for normality. (See Section 7.13.)

EXERCISES

Find, by coding, the arithmetic mean for each frequency distribution selected from Exercises 1–8.

1. Salaries of employees. (Ex. 3.5.)

2. Brain-weights of Swedish males. (Ex. 3.6.)

3. Ages of admirals. (Ex. 3.7.)

4. I.Q.'s of runaway boys. (Ex. 3.8.)

5. Weights of ampuls. (Ex. 3.11.)

6. Thicknesses of washers. (Ex. 3.12.)

7. Earnings of negro workers. (Ex. 3.13.)

8. Numbers of correct answers. (Ex. 3.14.)

9. What factors make the value of the arithmetic mean daily wage determined from the frequency distribution, Ex. 3.10, open to criticism?

10. Compute the arithmetic mean for the data of Ex. 3.16. Note the reversed order of class intervals.

11. The frequency distribution of Ex. 3.13 is very skew. Would the arithmetic mean computed from it be inaccurate because of this fact?

Find the standard deviations for the following distributions.

12. Ex. 1.   13. Ex. 2.   14. Ex. 3.

15. Ex. 4.   16. Ex. 5.   17. Ex. 6.

18. What would you say about the usefulness of any measure of dispersion computed from a J-distribution? Assume that all grouping errors have been properly eliminated.

Construct a cumulative frequency table and compute the median for each of the frequency distributions indicated below. Use the method of interpolation with a cumulative frequency diagram when directed to do so by your instructor.

19. Ex. 1.   20. Ex. 3.   21. Ex. 4.

22. Ex. 5.   23. Ex. 7.   24. Ex. 8.

25. Weight, per mille, of United States Army troops. (Ex. 3.15.) Compare the median weights at mobilization and at demobilization.

26. Construct on the same scale the two cumulative frequency diagrams for the data of Exercise 3.15. Compare them and draw conclusions.

27. Suppose that the cumulative frequency diagram for a given frequency distribution were a straight line. What would this indicate?

Compute (a) Q_1, (b) Q_3, (c) Q, (d) the 90 percentile, (e) the 35 percentile, (f) the 60 percentile, (g) V, for the distributions indicated below.

28. Ex. 1.   **29.** Ex. 3.   **30.** Ex. 4.

31. Find the modal class of each of the distributions of Exercises 3.5–12, 16.

32. Find the percentile corresponding to each of the following:
(a) A salary of $8000 for the salaries of Ex. 3.5.
(b) An I.Q. of 97 for the runaway boys of Ex. 3.8.
(c) A weight of 93 mg for the ampuls of Ex. 3.11.
(d) 20 answers correct, Ex. 3.14.
(e) 58 items correct, Ex. 3.16.

33. Assume that the mean weight of elephants is 6000 pounds and that the standard deviation is 1800 pounds. Assume that the corresponding statistics for man are 148 and 21 pounds. Which are more variable in weight, men or elephants?

34. The mean population of a group of 72 towns was 13,450 with $s = 4220$. The mean tax rate was $28.35 with $s = \$4.19$. Which was more variable, population or tax rate?

35. Which is more variable, heights or weights of B.U. women students, given for heights, $\bar{x} = 64.1$ inches and $s = 2.43$ inches, and for weights, $\bar{x} = 126$ pounds and $s = 18.1$ pounds?

36. For August the price relatives of foods showed a mean of 112 and a standard deviation of 8. The price relatives for clothing showed a mean of 124 and a standard deviation of 10. Which showed greater variability, the relatives for food or for clothing?

37. The pupils of two schools took the same verbal aptitude test with the following results.

	School A	School B
No.	153	161
$\bar{x}$	528	540
s	85	102
a_3	0.20	0.72

What information do you glean from these data?

38. The assessed valuations of the residential houses of two towns were tabulated. Their frequency distributions showed the following statistics:

	Millvale	Short Falls
No.	563	432
$\bar{x}$	$9462	$9035
s	$2112	$2817
a_3	0.18	1.10

Compare as completely as you can these towns with respect to valuation.

39. In a psychological experiment the numbers of seconds required for a group of adults to complete a certain operation were as follows:

26, 23, 25, 20, 27, 24, 24, 28, 32, 23.

(a) Find the range. (b) Find the mean deviation. (c) Find the standard deviation. (d) Estimate the standard deviation from your answer to (b). (e) Estimate the standard deviation from your answer to (a). (f) Which of these last two estimates do you consider the better one? Why?

40. For certain test scores the median is 70 and the quartile deviation is 10. What information does this statement yield?

41. During the months of June, July, and August the *median number* of pounds of butter produced per day at a certain dairy was 567 with a *quartile deviation* of 132. What does each italicized phrase mean with respect to these data?

42. The test scores for a large, fairly homogeneous group of students showed a median of 478 and a standard deviation of 82. What do you estimate to be the range of scores of the middle 50 per cent of this group?

43. Find the standard deviates corresponding to values in each of the following:
(a) A stature of 70 inches when the mean is 68 inches and the standard standard deviation, 2.5 inches.
(b) An electric lamp life of 570 hours when the mean is 950 hours and the standard deviation, 150 hours.
(c) A pulse rate of 86 beats per minute when the mean is 72 and the standard deviation, 5.

44. What information does each of the following statements yield?
(a) The median age of readers of College Examinations should be about 40 years.
(b) The quartile deviation of the wages of clerks in the A.B.C. department store is $10.54.
(c) Chris Black's score on the intelligence test corresponded to the 86th percentile.
(d) With a mean of 410 and a standard deviation of 80, the standard deviate was -0.5.
(e) The 35 percentile wage was $76.46.

45. What can you say about a frequency distribution with (a) $a_3 = 0.07$? (b) $a_3 = -1.6$? (c) $a_3 = 0.5$? (d) $a_3 = -0.20$? (e) $a_3 = 1.24$?

46. The grades of 240 pupils on a quiz are shown in the frequency table below. (a) Find Q_1, Q_2, and Q_3. (b) Between what limits do the middle 50 per cent of the grades lie? (c) Find the quartile deviation, Q. (d) What per cent of the grades lie within a distance Q, of the median? Find the answer by means of interpolation in the table. (e) Why is the answer to (d) not equal to 50 per cent?

Grade	Frequency
0–9	1
10–19	3
20–29	7
30–39	11
40–49	17
50–59	23
60–69	32
70–79	60
80–89	51
90–99	35
	240

PROBABILITY

5

5.1. Introduction

Many statements made in common parlance contain elements of uncertainty: "It will probably rain tomorrow." "The American League team is likely to win the World Series." "This new drug might be effective in combatting influenza." If you were to add to the first statement "I'll bet you two to one that it will rain," then you would be attempting, in a crude way, to measure your uncertainty. Statistics may be said to be concerned with "decision making in the face of uncertainty" and we may say, in a preliminary way, that mathematical probability is a measure of our uncertainty. This description is, however, inadequate for our purposes, so we shall proceed to define probability in a more precise manner.

5.2. Experiments, Events, and Sample Space

Statistics deals with data that are obtained from experiments or observations, either actual or imagined. We may record the pulse rates of hockey players, we may examine the grades of students in a calculus class, or we may imagine what would happen if two pennies were tossed. These are examples of statistical experiments. The results or outcomes of such experiments are called events and each is often symbolized by the letter x or E. Since such

87

events are the results of chance and vary from observation to observation or from experiment to experiment, we shall employ the symbol x for such random or chance variables (Section 1.5).

The pulse rates x, of 10 hockey players might have been found to be the set of numbers (of beats per minute)

$$68, 72, 65, 69, 70, 70, 74, 72, 69, 69.$$

Each of these numbers represents a *simple event*, or observation. Moreover, these might be interpreted as points on a line (Figure 5–1). Note that the point, 69, is counted three times, 70, twice, and 72, twice. If we were interested

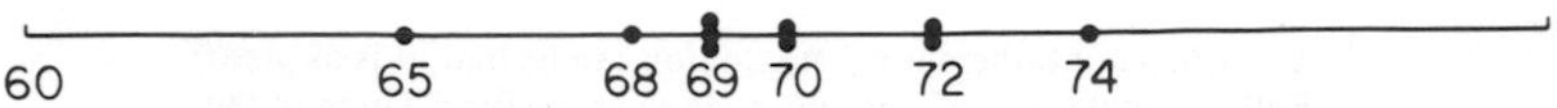

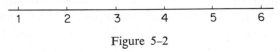

60   65   68 69 70   72   74

Figure 5–1

in the number of pulse rates under 72, then the outcome, $x < 72$, would constitute a *compound event* made up of seven simple events, the seven observations, 68, 65, 69, 70, 70, 69, and 69. These are the seven points to the left of the point, 72. These numbers or points constitute a *subset* of the observed ten numbers or points.

Let us take another illustration. If we were to toss a die, the numbers that could appear are 1, 2, 3, 4, 5, or 6. Again, this set of numbers could be represented by points on a line (Figure 5–2). Each point denotes a simple event. Such a configuration of points representing *all possible simple events*

1   2   3   4   5   6

Figure 5–2

is called an event space or, more commonly, a *sample space*. The outcome, "an odd number," is a compound event, and is represented by the subset of points, 1, 3, and 5. Intuitively one feels that if the die is an honest one, repeated tossings would produce these six possible outcomes about the same number of times. The sample space here is one-dimensional.

A third illustration arises if we make a random toss of two pennies and observe each *pair* of outcomes. Four simple events may occur: they may both show tails, one may show tail and the other head, the other may show head and the one tail, or both may show heads. These four simple events may be symbolized by *TT*, *TH*, *HT*, and *HH* respectively. Another useful method of designating them is by means of the four pairs of digits: (0, 0), (0, 1), (1, 0), (1, 1), where the first digit represents the number of heads appearing on penny A and the second represents the number of heads appearing on penny B. These pairs of numbers may be plotted as points (x, y) on a plane (Figure 5–3). Such a configuration or set of points is said to constitute a sample space for the experiment of tossing two pennies.

These four points represent all possible outcomes and the sample space is two-dimensional.

As an example of a compound event we may consider the possibilities that the two pennies show differently. This compound event, "the appearance

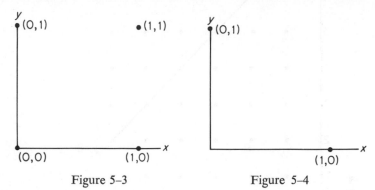

Figure 5–3   Figure 5–4

of different faces" may occur in two ways, by means of the two simple events, *TH* or (0, 1) and *HT* or (1, 0). The compound event is thus represented by a portion or subset of the sample space as shown in Figure 5–4.

Consider next the experiment of throwing a pair of dice; say one is black, the other, white. Note that the difference in color merely emphasizes the fact that they are different dice. There will be 36 simple events possible since the black die may show any face numbered from 1 to 6, and for each of these six possibilities the white die may also show any number from 1 to 6. These simple events are (1, 1), (1, 2), (1, 3), …, (1, 6), (2, 1), (2, 2), (2, 3), …, (2, 6), …, (6, 1), (6, 2),…, (6, 6). The totality of such possible outcomes may be represented by the sample space of Figure 5–5, consisting of the lattice of 6×6 or 36 points. As an example of a compound event for a pair of dice we consider "the sum 8." This sum may be obtained in the following ways and in no other way:

Black die	White die
x	y
2	6
6	2
3	5
5	3
4	4

Thus there are 5 simple events (points) constituting the compound event, "the sum 8;" the subset of points lying on the diagonal from (2, 6) to (6, 2) represent the compound event, $x + y = 8$.

As another illustration we may consider first the compound event "an even sum" ($x + y =$ an even number). This is represented by the larger

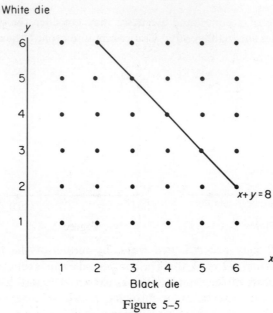

Figure 5–5

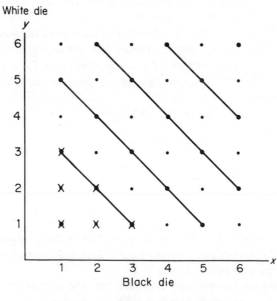

Figure 5–6

dots in Figure 5–6 lying on the alternate diagonals extending downward from left to right. Secondly, "a sum less than 5" ($x + y < 5$) is pictured by means of the 6 crossed dots in the lower left hand corner. Finally, the compound event, "an even number sum less than 5" is shown by means of the dots both large and crossed. In the language of sets, these dots constitute the *intersection* of two subsets.

5.3. Discrete and Continuous Space

In statistics we study variables such as heights of men, scholastic aptitude scores, numbers of children in families, sizes of men's socks, lengths of life of electric light bulbs, yields of corn, and so on. Some of these quantities can vary only by finite increments, by observable jumps in value. Two families can differ in numbers of children by not less than 1. We can have 0, 1, 2, 3, ... children. Men's socks are manufactured only in one-half sizes; the next size sock is either $\frac{1}{2}$ smaller or $\frac{1}{2}$ greater. These are examples of *discrete* variables. On the other hand, the stature of a man increases gradually, by infinitesimal amounts. Between any two given heights any number of heights is possible, although no measuring apparatus is capable of distinguishing among all of them. A similar property is characteristic of lives of electric lamps. These are example of *continuous* variables.

Let us examine briefly a sample space for a continuous variable. Let x

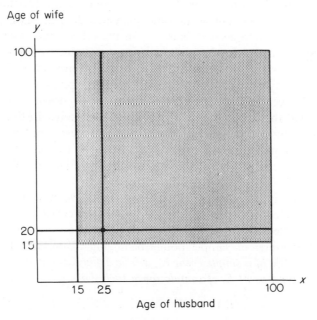

Figure 5–7

stand for the number of years of age of a husband. Let us assume that there
are no husbands existent younger than 15 nor older than 100 years. Then
the sample space for *ages of husbands* may be represented by all points on
the *x*-axis from 15 to 100. If we make a similar convention for *ages of wives*
(points on the *y*-axis from 15 to 100) then the sample space for *ages of married
couples* consists of all points within the rectangle bounded by the two vertical
lines at $x = 15$ and $x = 100$, and by the two horizontal lines at $y = 15$ and
$y = 100$. The simple event for a couple, "husband of age 25 and wife of age
20" is represented by the single point (25, 20). The compound event "married
couples with husbands older than 25" is represented by the subset of points
of the sample space to the right of the vertical line through $x = 25$. The
compound event "married couples with husbands older than 25 and wives
older than 20" is represented by the subset of points in the upper right hand
rectangle of the sample space.

5.4. Probability

If we return to the sample space for the tossing of two pennies (Figure
5–3), we sense intuitively that any one of the four possible outcomes *TT, TH,
HT, HH*, is as likely to happen as any other and therefore it seems reasonable
to assign the probability number $\frac{1}{4}$ to each of these outcomes. In fact, if we
were to toss a pair of pennies a large number of times, say, 1000, we would
find that each of the four possible events would occur roughly about 250
times or about 25 per cent of the time, provided the pennies were not badly
worn, mutilated, or otherwise biased. On the other hand, when the author
repeatedly tossed a pair of thumb tacks on to a table top and recorded the
number of occurrences of each of the four possible outcomes, *DD, DU, UD,
UU*, where *D* symbolizes "point down" and *U*, "point up" he found that
"point up" occurred less frequently than "point down" so that he felt
confident that these four events were not equally likely. In this case to each
possible outcome or sample point we could not reasonably assign the same
probability number. In an experiment with 100 tosses the author made
preliminary estimates of probabilities as follows: $P(DD) = 0.34$, $P(DU) =$
0.25, $P(UD) = 0.21$, $P(UU) = 0.20$.

In view of the foregoing remarks we shall now more formally *assume*
that every conceivable outcome or simple event corresponds to just one
point in sample space and that every compound event corresponds to an
aggregate or *subset* of sample points belonging to the sample space. *The
probability $P(E_i)$, of a simple event E_i, will be defined as a non-negative number
associated with a point E_i of sample space such that*

$$P(E_1) + P(E_2) + \cdots + P(E_n) = 1,$$

where the total number of points in the sample space is *n*.

Thus for the toss of two pennies

$$P(TT) + P(TH) + P(HT) + P(HH) = \frac{1}{4} + \frac{1}{4} + \frac{1}{4} + \frac{1}{4} = 1:$$

For the toss of two thumb tacks

$$P(DD) + P(DU) + P(UD) + P(UU) = 0.34 + 0.25 + 0.21 + 0.20$$
$$= 1.$$

Here the four probabilities are suggested by the author's experiment but cannot be regarded as accurate.

The extension of the preceding definition of probability to include continuous sample spaces where n becomes infinite cannot be discussed in a book of this level.

In many cases a probability number cannot be assigned to a simple event by any process of inspection or geometric analysis, as we do in the case of a penny or a die. Instead we assume that if N is the number of repetitions made under identical conditions, and f, the frequency of occurrence of an event E, then for large N, the ratio f/N is a good approximation to $P(E)$. The exact probability would then be formally defined as the limit which f/N is assumed to approach as N becomes infinite. In symbols we write

$$P(E) = \lim \frac{f}{N}.$$

The vast field of insurance—life, accident, casualty, etc.—is built upon probabilities defined in just this manner.

Examples of "probability numbers" assigned to simple events follow:

(a) The probability that a penny shows head is $\frac{1}{2}$.

(b) The probability that two pennies both show heads is $\frac{1}{4}$.

(c) The probability that a pair of dice, one black, the other white, shows 3 on the black die and 5 on the white, is $\frac{1}{36}$.

The reader may wish to refer to the appropriate space diagrams in order to see the reasons for defining the probabilities in (b) and (c).

If a compound event A, consists of the simple events $E_1, E_2, ..., E_k$, we define its probability $P(A)$, by means of the equation

$$P(A) = P(E_1) + P(E_2) + \cdots + P(E_k).$$

We cite the following examples of this definition.

(d) For the toss of two pennies the probability that they show the same faces, $P = P(HH) + P(TT) = \frac{1}{4} + \frac{1}{4} = \frac{1}{2}$.

(e) In the toss of a pair of dice the probability of obtaining the sum 6,

$$P(x = 6) = P(1, 5) + P(5, 1) + P(2, 4) + P(4, 2) + P(3, 3)$$
$$= \frac{1}{36} + \frac{1}{36} + \frac{1}{36} + \frac{1}{36} + \frac{1}{36}$$
$$= \frac{5}{36}.$$

(f) For a pair of dice the probability of obtaining a sum less than 13,

$$P(x+y<13) = P(x+y=2) + P(x+y=3) + \cdots + P(x+y=12)$$
$$= P(1,1) + [P(1,2) + P(2,1)]$$
$$+ [P(1,3) + P(3,1) + P(2,2)] + \cdots + P(6,6) = 1,$$

and the probability of obtaining a sum 13,

(g) $$P(x+y=13) = 0.$$

It is evident that mathematical probability is a number restricted to values from 0 to 1 inclusive. In symbols

$$0 \leq P(A) \leq 1.$$

In (f) the probability, 1, is associated with certainty—a sum less than 13 *must* occur when a pair of dice is thrown. In (g) the probability of an impossible event is zero.

5.5. The Addition Theorem

Let $P(A)$ and $P(B)$ be the probabilities for two events, A and B. Let $P(A + B)$ denote the probability that either A or B occurs or both occur. Let the events A and B be represented by points of the two regions marked A and B within the sample space S, as shown in Figure 5–8. These two

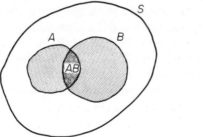
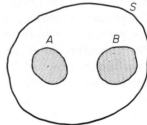

Figure 5–8   Figure 5–9

regions we shall assume have a common portion or *intersection AB*. To each point is associated a probability. The sum of the probabilities for the points in A yield $P(A)$, and the sum for the points in B, $P(B)$. Then

$$P(A + B) = P(A) + P(B) - P(AB), \qquad (5.1)$$

where $P(AB)$ designates the probability for both A and B to occur, for the addition of the probabilities $P(A)$ and $P(B)$ means that in adding all the points of A to all the points of B, the points of AB are counted twice and hence

$P(AB)$ must be subtracted from the sum $P(A) + P(B)$. Equation (5.1) states the *addition theorem* of probability. This theorem may be called the "either-or" theorem.

As an example of this law let us refer back to Section 5.2 and Figure 5–6 in order to calculate the probability that in the toss of two dice we obtain either *an even sum* or *a sum less than* 5.

$$P(A) = P(x + y = \text{an even number}) = {}^{18}\!/_{36} = \tfrac{1}{2},$$

because the 18 larger points in the sample space correspond to an even sum and each point has a probability of $\tfrac{1}{36}$ attached to it.

$$P(B) = P(x + y < 5) = {}^{6}\!/_{36} = \tfrac{1}{6},$$

because the 6 crossed points correspond to sums less than 5.

$$P(AB) = {}^{4}\!/_{36} = \tfrac{1}{9},$$

because the points both crossed and large correspond to the simultaneous occurrence of A and B. By the addition theorem (5.1)

$$P(A + B) = {}^{18}\!/_{36} + {}^{6}\!/_{36} - {}^{4}\!/_{36} = {}^{20}\!/_{36} = \tfrac{5}{9}.$$

If A and B exclude each other, that is, if they cannot occur simultaneously, they are termed *mutually exclusive* events and $P(AB) = 0$. In this case the addition theorem reduces to

$$P(A + B) = P(A) + P(B). \tag{5.2}$$

Here the two spaces A and B have no points in common (Figure 5–9). The symbol AB (in set theory, $A \cap B$) denotes the intersection of the two point sets, A and B; the symbol $A + B$ (in set theory $A \cup B$) denotes the union or sum of two sets. If $P(AB) = 0$, then the set AB is said to be an *empty* or *null* set. If the intersection of two sets is the null set, they are said to be disjoint.

5.6. The Multiplication Theorem

For the experiment of tossing a pair of dice we represented the sample space conveniently by means of a square array of 36 points (Figure 5–6). Suppose now that we ask, "If a pair of dice shows an even sum, what is the probability that this sum is less than 5?" We are now restricting the sample space to a subset of points corresponding to even sums only (18 such points) and asking, "Which of these possible points (outcomes) represent a sum less than 5?" There are 4 of these points out of the 18 and all have equal probabilities, hence $P = \tfrac{1}{18} + \tfrac{1}{18} + \tfrac{1}{18} + \tfrac{1}{18} = \tfrac{2}{9}$. Note that we have set down the condition that the sum, $x + y$, is even (event A), and then requested the probability for $x + y$ to be less than 5 (event B). This *conditional probability* is symbolized by the formula $P(B/A)$, which may be read "P of B line A" or in words, "the probability for the event B under the condition that event A has occurred." Here A is the event "$x + y$ is even" and B is the event "$x + y < 5$." The probability $\frac{2}{9}$, above, was obtained by dividing the number, $N(AB)$, of points in the subset "$x + y$ is even and less than 5" by the number, $N(A)$, of points in the subset "$x + y$ is even" so that

$$P(B \mid A) = \frac{N(AB)}{N(A)}$$

$$= \frac{N(AB)/N}{N(A)/N} = \frac{P(AB)}{P(A)} \tag{5.3}$$

where N is the number of points in the entire sample space. Of course, $N(A)$ is assumed not to be 0.[1]

In the practical applications of (5.3) it is usually $P(AB)$ that is difficult, if not impossible, to calculate so that a very useful form is obtained by clearing the equation of fractions. We have, then, the important *multiplication theorem*

$$P(AB) = P(A)P(B \mid A). \tag{5.4}$$

Stated in words this reads "The probability for the simultaneous occurrence of the events A and B equals the probability that A occurs multiplied by the probability that B occurs provided A has already occurred."

EXAMPLE 1.   Let us consider an urn containing 5 balls, alike in every observable respect save color. If 3 of these balls are white and 2 are black and we draw at random two balls from this urn without replacement, what is the probability that the first ball drawn is white and the second, black? Theorem (5.4) may be written

$$P(WB) = P(W)P(B \mid W).$$

The probability $P(W)$ that the first ball drawn is white is $\frac{3}{5}$. If the first ball drawn is white then 4 balls are left of which 2 are black; hence the probability, $P(B \mid W)$, that the second ball is black provided the first is white is $\frac{2}{4}$. Then

$$P(WB) = (\tfrac{3}{5})(\tfrac{2}{4}) = \tfrac{3}{10}.$$

The reader should verify that $P(BW) = \frac{3}{10}$ also. Since WB and BW are mutually exclusive events, the probability that the two balls drawn are of different color is $P(WB) + P(BW)$. This sum equals $\frac{3}{10} + \frac{3}{10} = \frac{3}{5}$.

EXAMPLE 2.   Suppose that we have two boxes. In Box No. 1 there are five sealed envelopes; 3 of them contain $1 bills, and two of them, $5 bills. In Box No. 2 there are ten sealed envelopes, 7 of them contain $1 bills, and 3, $5 bills. If a box is selected at random and from it an envelope is chosen, what is the probability that it contains a $5 bill?

The chance of selecting Box No. 1 is $\frac{1}{2}$, and if this is done, the probability of getting a $5 bill is $\frac{2}{5}$. Then by (5.4)

$$P_1 = \tfrac{1}{2} \cdot \tfrac{2}{5} = \tfrac{1}{5}.$$

[1]Although formula (5.3) was derived from a special case, we shall accept it as a definition of the conditional probability $P(B/A)$.

Similarly for Box No. 2

$$P_2 = \tfrac{1}{2} \cdot \tfrac{3}{10} = \tfrac{3}{20}.$$

Since P_1 and P_2 are the probabilities for two mutually exclusive events, the probability for either one event or the other to occur is by the addition theorem

$$P_1 + P_2 = \tfrac{1}{5} + \tfrac{3}{20} = \tfrac{7}{20}.$$

If the occurrence of either the event A or the event B is not influenced or conditioned by the other event, the two events are said to be *mutually independent*, hence

$$P(B \mid A) = P(B)$$

or

$$P(A \mid B) = P(A).$$

If such is the case then the multiplication theorem (5.4), for mutually independent events, becomes

$$P(AB) = P(A)P(B). \tag{5.5}$$

For example, the probability that, in two tosses of a penny, both show head

$$P(HH) = P(H)P(H)$$
$$= \tfrac{1}{2} \cdot \tfrac{1}{2} = \tfrac{1}{4},$$

for the result of one toss does not influence the result of the other toss.

5.7. Expected Value

In Section 4–1 we defined the arithmetic mean of a frequency distribution as

$$\bar{x} = \frac{1}{N} \sum_{i=1}^{n} f_i x_i, \tag{5.6}$$

where n is the number of classes and N, the total frequency, $\sum_{i=1}^{n} f_i$. We may rewrite this formula as

$$\bar{x} = \sum_{i=1}^{n} \left(\frac{f_i}{N} \right) x_i,$$

where f_i/N is the relative frequency of the ith class, which can be interpreted as a probability, p_i. Then

$$\bar{x} = \sum_{i=1}^{n} p_i x_i. \tag{5.7}$$

Note that $\bar{x}$ is the mean of a sample and p_i is an empirical probability, derived from the sample itself. These considerations lead us to the important concept of the *expected value* or *mathematical expectation*. Its formal definition follows:

If p_i is the probability of obtaining the value x_i, the expected value

$$E(x) = \sum_{i=1}^{n} p_i x_i \qquad (5.8)$$

where $\sum_{i=1}^{n} p_i = 1$.

The theoretical mean μ, of a discrete chance variable x, is defined as $E(x)$, where now p_i is a theoretical probability.

Example: Suppose that a box contains 20 small packages, all indistinguishable except for the contents. Assume that 10 of these packages each contains a cent, 5 packages each contains a nickel, 3 each a dime, and 2 each, a quarter. If a package is drawn at random, what is the expected value of the coin contained in it?

The probability, p_1, of drawing a cent is $^{10}\!/_{20}$ or 0.50; of drawing a nickel, $p_2 = ^{5}\!/_{20} = 0.25$; of a dime, $p_3 = ^{3}\!/_{20} = 0.15$; and of a quarter, $p_4 = ^{2}\!/_{20} = 0.10$. Note that $p_1 + p_2 + p_3 + p_4 = 1$ and that $x_1 = 1$, $x_2 = 5$, $x_3 = 10$, and $x_4 = 25$.

$$\begin{aligned} E(x) &= \Sigma \, p_i x_i \\ &= (0.50 \times 1) + (0.25 \times 5) + (0.15 \times 10) + (0.10 \times 25) \\ &= 5.75. \end{aligned}$$

The answer, 5.75 cents, may be interpreted as follows: If the selection of a package were repeated a large number of times under the same conditions, the mean value of the amounts obtained would approximate 5.75 cents.

The theoretical variance σ^2, of a discrete chance variable x, is defined as

$$E(x - \mu)^2 = \sum_{i=1}^{n} p_i (x_i - \mu)^2 \qquad (5.9)$$

where $\mu = E(x)$.

EXERCISES

1. A die has three faces numbered 1, two faces numbered 2, and one face numbered 3. Make a diagram of the sample space for the possible outcomes when the die is thrown. Label each point with its appropriate probability.

2. Two dice have the same color on opposite faces of each and the colors red, white, and blue appear on each die. (a) Construct a sample space diagram of the possible color outcomes when the pair of dice are thrown. (b) What is the probability that they show different colors? (c) What is the probability that white does not show?

3. Refer to Exercise 3.5 and find the probability that an employee chosen at random from the group has a salary (a) between $5000 and $6000; (b) greater than $7000.

4. Refer to Exercise 3.16 and (a) calculate the probability that a student chosen at random had more than 67 items correct. (b) Which would be more probable, that this student had more than 55 items correct or that he had fewer than 32 items correct?

5. What is the probability that the throw of two dice is under 7?

6. If a pair of dice is tossed what is the probability that the sum is either 7 or 11?

7. What is the probability of obtaining the sum 11 when 3 dice are tossed?

8. Of 150 patients examined at a clinic it was found that 90 had heart trouble, 50 had diabetes, and 30 had both diseases. What percentage of the patients had either heart trouble or diabetes?

9. A history class consisted of 36 men and 24 women. Twenty-eight of the men and 20 of the women received grades of C or better. If a student were chosen at random from this class, what is the probability that this student has either a grade of C or better, or is a woman?

10. If a pack of cards is cut what is the probability that it shows an ace, a king, a queen, a jack, or a diamond?

11. In a certain college for men 5 per cent of the senior class were elected to Phi Beta Kappa, 10 per cent of the class were veterans, and 10 per cent of the veterans were elected to Phi Beta Kappa. What is the probability that a senior chosen at random was either a veteran or a Phi Beta Kappa man?

12. A box contains 3 red balls and 8 black balls, all of the same size and material. If two balls are drawn in succession without replacement, what is the probability that both are red?

13. From a bag containing 5 black and 3 white balls, 3 balls are drawn in succession without replacements. What is the probability that all 3 are black?

14. A box contains 5 red balls, 10 white balls, and 15 blue balls, all of the same size and material. If 3 balls are drawn in succession without replacements, what is the probability that they are of different color from one another?

15. If 60 per cent of American males of age 20 and 65 per cent of American females of age 20 live to be 70, what is the probability that an American couple married when they were 20 years old will live to celebrate their golden wedding?

16. In New England the distribution of the 4 basic blood groups has been found to be as follows: O, 45 per cent; A, 40 per cent; B, 10 per cent; AB, 5 per cent. What is the probability, for a random New England

couple, that (a) both are of type A? (b) neither is of type O? (c) the wife is of type A and the husband of type B? (d) one is of type A and the other of type B? (e) they are of different types?

17. If two guinea pigs, one of pure black race and the other of pure white race are mated, the probabilities that each offspring of the second generation is pure white, pure black, or of mixed color are respectively $\frac{1}{4}$, $\frac{1}{4}$, and $\frac{1}{2}$. What is the probability that 3 such offspring would possess different colors from one another?

18. As a rule 10 per cent of a certain type of Christmas tree bulb are defective. If a random sample of 10 such bulbs is selected from a lot and tested, what is the probability that none are defective? You need not compute the answer decimally.

19. In a large club there are twice as many women as men. If 3 committee members are to be chosen by lot, what is the probability that two are men and one is a woman? If the club consisted of only 9 members with the same ratio of sexes as before, what would your answer be?

20. If a die is thrown 3 times, what is the probability that (a) all throws show 6? (b) all throws are alike?

21. A box contains 5 white and 2 black balls. A second box identical with the first, contains 3 white and 5 black balls. One box is chosen and a ball withdrawn from it. What is the probability that this ball is white?

22. If A, B, and C are events, derive a formula for $P(A + B + C)$. This will be an extension of (5.1) and may be obtained with the aid of a diagram of the same kind as Figure 5–8.

23. A bag contains 7 pennies, 3 nickels, 2 dimes, 1 quarter, and a half dollar, the coins contained one each in uniform boxes. What is the expectation of a person drawing a box at random?

24. A lottery has one prize of $1000, two prizes of $500, five of $100, and fifty of $5. If 1000 tickets are sold what is the value of a ticket? Do this problem by two methods.

25. A pair of dice is thrown and the thrower is to receive a number of dollars equal to the sum that appears. What is his expectation?

26. Suppose that in general 20 per cent of the patients afflicted with a specific disease die. In a random sample of 3 such patients what is the probability for 2 deaths?

27. Approximately 10 per cent of type A radio tubes received by a radio set manufacturer are defective. Three tubes of type A are used in each set. With respect to the performance of these tubes, what proportion of sets will operate properly?

THE BINOMIAL
DISTRIBUTION

6

"And I believe that the Binomial Theorem and a Bach
Fugue are, in the long run, more important than all the
battles of history."
"I believe in the wisdom of often saying 'probably' and
'perhaps.'"
JAMES HILTON
This Week Magazine[1]

6.1. Introduction

It was stated in the opening chapter of this book that modern statistics
could trace its ancestry in part to that activity of questionable value, gambling.
The solution of a certain problem in a game of chance has indeed furnished
us with a powerful tool for studying frequency distributions of an important
kind. The fundamental theory is fairly simple if we eliminate the technical
details of rigorous proofs. It will be the aim of this chapter to study this
tool and its wide applications. To this end, we introduce certain preliminary
and fundamental notions upon which we may build an appreciation of the
role of probability in modern statistics. We begin with a brief survey of
permutations and *combinations*.

6.2. A Fundamental Assumption

The following statement will be taken as obviously true.

*If a thing can be done in **m** different ways, and if, after it is done in one of
these ways, a second thing can be done in **n** different ways, then the two things
together can be done in **mn** different ways in the order named.*

[1] Copyright, 1937, by the United Newspapers Magazine Corporation.

For example, if a student may elect any one of five courses at nine o'clock, and any one of three courses at ten o'clock, he may choose a combination of two courses in fifteen different ways.

6.3. Factorials

We define *factorial N*, symbolized by $N!$, as $N(N-1)(N-2) \cdots 3 \cdot 2 \cdot 1$ where N is assumed to be a positive integer. Thus, $4! = 4 \cdot 3 \cdot 2 \cdot 1 = 24$ and $7! = 7 \cdot 6 \cdot 5 \cdot 4 \cdot 3 \cdot 2 \cdot 1 = 5040$. It follows that:

$$4! = 4 \cdot 3!$$
$$3! = 3 \cdot 2!$$
$$2! = 2 \cdot 1!$$

If we carry this process forward in a mechanical manner, we should write

$$1! = 1 \cdot 0!$$

In order for the symbol $0!$ to have meaning, it would seem that we should let $0! = \dfrac{1!}{1} = 1$. We shall therefore define the value of this symbol $0!$ as equal to 1.

6.4. Permutations and Combinations

Each different arrangement of a group of things is called a *permutation*. A group of things considered without reference to their order within the group is called a *combination*.

Consider, for example, the following groups of letters: *abc, acb, bac, bca, cab*, and *cba*. Each of the six groups contains the same combination of letters, but each has a different arrangement from any other. There are represented, therefore, six different permutations of three letters, but only one combination of three.

On the other hand, if from the three letters *abc* we choose two letters at a time, we can make at most six permutations; namely, *ab, ba, ac, ca, bc*, and *cb*. But the groups *ab* and *ba* constitute the same combination of letters; likewise for *ac* and *ca*, and for *bc* and *cb*. Therefore, only three combinations are possible, taking only two at a time.

6.5. Fundamental Theorems on Permutations

The number of permutations possible of r things taken from N different things,

$$P(N, r) = \frac{N!}{(N-r)!} \cdot \tag{6.1}$$

Proof: We can choose the first thing in N ways; having chosen it in a given way, we can choose the second thing in the remaining $(N-1)$ ways. By the Fundamental Assumption, we can, then, choose the first two things in $N(N-1)$ ways. Having chosen the first two things, there are $(N-2)$ left from which to choose the third. Hence, by the Fundamental Assumption, we may choose the first three things in $N(N-1)(N-2)$ ways. Continuing in this manner until r things have been chosen, we find that they may be selected in

$$N(N-1)(N-2)\cdots(N-\overline{r-1})$$

ways. Since the next factor in order in this product would be $N-r$, we multiply this product by $\dfrac{(N-r)!}{(N-r)!}$ to obtain:

$$\frac{N(N-1)(N-2)\cdots(N-\overline{r-1})(N-r)!}{(N-r)!}.$$

But the numerator is equivalent to $N!$; hence,

$$P(N, r) = \frac{N!}{(N-r)!}.$$

Example. If five sprinters compete in the final race of a 100 yard dash, in how many different ways may three prizes be won?

Here the order in which they finish determines, of course, the first, second, and third prize awards. $N = 5$, $r = 3$; hence by (6.1)

$$P(5, 3) = \frac{5!}{2!} = 60.$$

Therefore the prizes may be won in 60 different ways.

As a corollary to this theorem, we state the following:

The number of permutations possible of N different things taken all at a time,

$$P(N, N) = N!. \tag{6.2}$$

The proof appears when we set $r = N$ in $(N-r)!$ to obtain $0!$ or 1 as the denominator in (6.1) above.

The number of permutations possible of N things consisting of k different groups such that n_1 things are alike in the first group, n_2 are alike in the second group, ..., and n_k are alike in the kth group where $n_1 + n_2 + \cdots + n_k = N$, is given by the formula

$$P(N:n_1, n_2, ..., n_k) = \frac{N!}{n_1! n_2! \cdots n_k!} \tag{6.3}$$

Proof: If the symbol on the left represents the desired number of permutations, and if each n_i things $(i = 1, 2, ..., k)$ were replaced by n_i things all different from one another and from all other things in the group of N, then the number of permutations of N different things taken all at a time could be obtained by multiplying $P(N: n_1, n_2, ..., n_k)$ by $n_1! \, n_2! \cdots n_k!$, for $n_i!$ represents the number of permutations possible from n_i different things taken all at a time. [Formula (6.2)]. Thus

$$P(N: n_1, n_2, ..., n_k) n_1! \, n_2! \cdots n_k! = N!$$

whence Formula (6.3) follows.

6.6. A Fundamental Theorem Concerning Combinations

The number of combinations possible of r things taken from N different things,

$$C(N, r) = \frac{N!}{(N - r)! \, r!}. \tag{6.4}$$

Proof: Any given combination of r different things is susceptible to $r!$ permutations, Formula (6.2); hence, if each combination is multiplied by $r!$, we obtain the total number of permutations, $P(N, r)$. Thus,

$$P(N, r) = C(N, r)r!;$$

whence, $$C(N, r) = \frac{P(N, r)}{r!} = \frac{N!}{(N - r)!r!}.$$

Example.   A committee of 5 is to be chosen from a club of 12 members. In how many ways may the committee be chosen?

Solution: Since a different order of a given group of 5 does not constitute a different committee, the problem is that of finding the number of combinations of 5 people at a time selected from 12. Using Formula (6.4), we have:

$$C(12, 5) = \frac{12!}{5!7!}$$
$$= 792.$$

6.7. Binomial Probability

Assume that an event can happen in two different ways, one called a success, the other, a failure. Let p be the probability for a success, q, the probability for a failure. Suppose, next, that an event is repeated N times under essentially the same conditions so that the outcome of each event is independent of the other outcomes. What is the probability for exactly x successes?

The probability that success results in each of the first x trials is, by an obvious extension of the multiplication theorem, p^x, [see (5.5)] and the probability that failure results in each of the remaining $(N - x)$ trials is, by the same theorem, q^{N-x}, hence the probability that the first x trials are successful and that the last $N - x$ trials are failures is $p^x q^{N-x}$. But this is the probability associated with one particular order of successes and failures, namely,

$$\underbrace{pp \cdots p}_{x \text{ factors}} \underbrace{qq \cdots q}_{N - x \text{ factors}}.$$

Exactly x successes in N trials can happen in any one of $\dfrac{N!}{x!(N - x)!}$ different orders [Formula (6.3)], hence, by the addition theorem for mutually exclusive events, [Formula (5.2)], the probability for x successes in N trials is

$$P(x) = \frac{N!}{x!(N - x)!} p^x q^{N-x}. \tag{6.5}$$

EXAMPLE 1. The probability that two heads appear when six pennies are tossed is:

$$\frac{6!}{2!\,4!} \left(\frac{1}{2}\right)^2 \left(\frac{1}{2}\right)^4 = \frac{15}{64},$$

for $p = q = \frac{1}{2}$, $N = 6$, and $x = 2$. Note that the tossing of six pennies simultaneously is experimentally equivalent to the tossing of one penny six times in succession.

EXAMPLE 2. A box contains 10 balls alike in every respect save color. Seven are white and three are black. A blindfolded person selects a ball at random and the color is then noted. Then the ball is replaced and all balls thoroughly mixed again. A second ball is withdrawn and followed by the same procedure, until 5 drawings have been made. What is the probability that three balls drawn were white and two black?

The probability, p, for drawing a white ball is $\frac{7}{10}$; the probability, q, of drawing a black one is $\frac{3}{10}$. The number of trials, N, is 5, and $x = 3$, Then the desired probability is:

$$\frac{5!}{3!\,2!} \left(\frac{7}{10}\right)^3 \left(\frac{3}{10}\right)^2 = 0.3087.$$

6.8. A Decision Problem

Suppose that we possess a well-worn penny and wonder if it is biased, that is, if it tends to show heads more frequently than tails, or vice versa. We may satisfy our curiosity in the following manner. Let us set up the hypothesis that the coin is a true one, that is, unbiased. In more precise mathematical language we may state our hypothesis, H_0, called the *null* hypothesis, as $p = \frac{1}{2}$, where p is the probability for head. Then let us toss this penny 10 times and observe the number of times a head appears. If head appears too often or too seldom we shall reject the hypothesis, H_0, and thus decide that the coin is biased; otherwise we shall decide that the penny is a fair one.

Before continuing this discussion we must make clear that our decision is to be based on just one experiment consisting of ten tosses of the coin. Of course we might increase the number of tosses to 50, or 100, or even more. This would be easy to do in the case of a coin, but in many practical problems the number of observations possible may be restricted. For example, suppose that we are testing the hypothesis that Drug Q produces an undesirable after-effect in 50 per cent of the patients to whom it is administered and we have only 10 such patients available. Our procedure for the immediate future might be governed by the outcome from this sample of ten. Here we wish to decide what number of bad after-effects will cause us to abandon the drug in favor of another. In all phases of human activity decisions have to be made in the face of incomplete evidence.

Intuitively we feel that a well-behaved penny should show head about half of the time, although we realize that results sometimes are erratic or unusual. Ordinarily, then, we expect somewhere near 5 heads to show; a deviation say, of two heads from the "expected" 5 should not cause us surprise, therefore, 3, 4, 5, 6, or 7 heads might well be considered to be consistent with the hypothesis that the coin is a fair one. Where should we draw the line between results that cause us to accept our hypothesis and those that compel us to reject it, particularly when we realize that an unbiased coin may, at times, give surprising results? These questions lead us to the fundamental problem of "making a decision in the face of uncertainty." We cannot, positively, with *any* number of tosses of the coin make certain that our decision is the correct one, but we *can* calculate the probability of a decision being wrong. Our problem, restated, is to agree upon the numbers of heads appearing that seem abnormal to us and therefore cause us to reject the hypothesis that the coin is true. All other numbers of heads will be considered to be consistent with this hypothesis.

Let us agree that if 0, 1, 9, or 10 heads appear, we shall consider the null hypothesis, H_0, to be refuted. Zero or one head (i.e. 10 or 9 tails) will be interpreted as evidence that the probability for head, $p < \frac{1}{2}$, that is, that the penny is biased in favor of tails, and 9 or 10 heads (i.e. 1 or 0 tail) that $p > \frac{1}{2}$, or that the penny is biased in favor of heads. Any other number of heads 2, 3, 4, 5, 6, 7, or 8, will be construed as consistent with H_0.

6.9. Rejection and Acceptance Regions

The probabilities for 0, 1, 2, ..., 10 heads for an unbiased penny may be found from Formula (6.5) by letting $N = 10$, $p = q = \frac{1}{2}$, and $x = 0, 1, 2,$ ..., 10, respectively. The 11 probabilities are computed to 4 decimal places as shown in Table 6–1, and are represented by the lengths of the vertical lines in Figure 6–1.

In Figure 6–1 the 11 points on the x-axis corresponding to numbers of heads, form a set of distinct points, a one-dimensional sample space. The

Table 6–1

No. heads, x	Probability, $P(x)$
0	$^{1}/_{1024} - 0.0010$
1	$^{10}/_{1024} = 0.0098$
2	$^{45}/_{1024} = 0.0439$
3	$^{120}/_{1024} = 0.1172$
4	$^{210}/_{1024} = 0.2051$
5	$^{252}/_{1024} = 0.2461$
6	$^{210}/_{1024} = 0.2051$
7	$^{120}/_{1024} = 0.1172$
8	$^{45}/_{1024} = 0.0439$
9	$^{10}/_{1024} = 0.0098$
10	$^{1}/_{1024} - 0.0010$

two points at each end of the set, that is, the points $x = 0, 1, 9,$ and 10 form a *critical* or *rejection region* of the sample space for the hypothesis, $p = \frac{1}{2}$. Since these points constitute two ends or tails of the distribution diagram of Figure 6–1, we have a *two-tail critical region*, or we have a *two-tail test* of our hypothesis; either too many or too few heads cause us to reject H_0. The

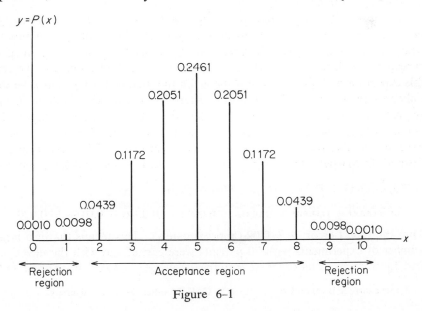

Figure 6–1

points $x = 2, 3, 4, 5, 6, 7$, and 8 constitute the *acceptance region*. An outcome that yields a point in the latter region causes us to accept the hypothesis; an outcome in the former region causes us to reject it.

6.10. The Type I Error

The probability of obtaining one or another of the four outcomes 0, 1, 9, 10 heads, when the coin is true is found from Table 6–1 or Figure 6–1 to be the sum of their separate probabilities:

$$0.0010 + 0.0098 + 0.0098 + 0.0010 = 0.0216.$$

This may be interpreted to mean that in repeated experiments of throwing a true coin 10 times, 0, 1, 9, or 10 heads will appear 2.16 per cent of the time. Thus whenever we reject the hypothesis that the coin is true because 0, 1, 9, or 10 heads show, we have a probability of 0.0216 of being wrong, or stated otherwise, about 2.16 per cent of the time that we make decisions on this basis, we shall be making the wrong decision. This is the risk that we run. On the other hand, $1 - 0.0216$ or 0.9784 is the probability of making the correct decision, "the coin is true," whenever we get 2, 3, 4, 5, 6, 7, or 8 heads, which correspond to points in the acceptance region of our sample space.

The probability of rejecting the null hypothesis when it is true is designated by the Greek letter alpha, α, and the error thus committed is called a type I error.

Above $\alpha = 0.0216$.

Example:[2] Suppose that past experience has shown that Drug Q produces an undesirable after-effect in about 50 per cent of the patients to whom it is administered. A modification in the formula of the drug is expected to reduce the per cent of patients having bad after-effects. To test this expectation 20 patients are to be given the modified drug. Let p be the unknown probability for undesirable after-effects from the modified drug, so that we test the hypothesis, H_0, that $p = \frac{1}{2}$ (or greater) against the alternative hypothesis, H_1, that $p < \frac{1}{2}$. Let us agree that if 5 or fewer patients have undesirable reactions we shall reject H_0 in favor of H_1. By means of Formula (6.5) where $N = 20$, $p = \frac{1}{2}$, and $x = 0, 1, 2, 3, 4$, and 5, we find that

$$P(0) + P(1) + P(2) + P(3) + P(4) + P(5)$$

$$= 0.0000 + 0.0000 + 0.0002 + 0.0011 + 0.0046 + 0.0148 = 0.0207.$$

Thus the type I error, $\alpha = 0.02$ (approximately). This means that if a large number of experiments with 20 patients were carried out with the new drug and if it were no better or worse than the old drug ($p = \frac{1}{2}$), 5 or fewer patients

[2] This example is taken from a paper by G. E. Noether, Boston University.

would have adverse effects about 2 per cent of the time, and 2 per cent of the time we would be deciding that the modified drug is better, an incorrect decision. Note here that we are using a one-tail test, for only results that yield fewer after-effects than expected would cause us to reject the hypothesis, $p \geq \frac{1}{2}$ in favor of the alternative, $p < \frac{1}{2}$.

6.11. The Type II Error

The consequence of rejecting a hypothesis when it is true may be costly in terms of dollars and cents, or perhaps, in terms of health or even life. Likewise the acceptance of a hypothesis when it is false may have undesirable results.

The probability of accepting the null hypothesis when it is false is designated by the Greek letter beta, β, and the error thus committed is called a type II error.

Suppose that the true, but unknown probability of an after-effect from the modified formula were 0.30. For this value of p what is the probability that the number of after-effects exceeds 5, so that we would accept the hypothesis, $p = \frac{1}{2}$ when it is false? Here $N = 20$, $p = 0.30$, and $x > 5$. Then

$$P(6) + P(7) + \cdots + P(20) = 1 - [P(0) + P(1) + \cdots + P(5)]$$

$$= 1 - (0.0008 + 0.0068 + 0.0278 + 0.0716 + 0.1304 + 0.1789)$$

$$= 1 - 0.4163 = 0.5837.$$

Thus if $p = 0.30$ we would accept the hypothesis, $p = 0.50$, about 58 per cent of the time: our type II error, $\beta = 0.58$. Thus if we set $\alpha = 0.02$ (approximately) we run the risk 58 per cent of the time of not discovering that the modified drug is better if $p = 0.30$. Of course we should like to reduce the size of β but we can only do this at the expense of α. The smaller α is, the larger β becomes and vice versa. In a practical problem the experimenter must decide in advance what the consequences of a wrong decision are.

The possible decisions and the nature of the errors are shown in the table below.

Table 6–2

		Hypothesis is correct	false
Hypothesis	accepted	Decision is correct	Type II error
	rejected	Type I error	Decision is correct

6.12. Remarks

(1) In the drug experiment it should be noted that although the hypothesis to be tested was $p \geqq \frac{1}{2}$, we used only the value $p = \frac{1}{2}$ in order to determine the probability associated with the rejection region. If $p > \frac{1}{2}$ the chance for the number of bad effects to be 5 or fewer would be even smaller than if p equalled $\frac{1}{2}$, so that the rejection of the hypothesis would be more emphatic, or, to state the fact in a different way, the type I error would be even smaller than 0.02.

(2) In the coin-tossing experiment we were testing to see if the coin was biased—in what direction, we did not know, so we used a two-tail test. Either an excessively small number of heads or an excessively large number was taken as an indication of bias. In the drug experiment we hoped for a reduction in the customary number of bad effects so that only an abnormally small number of them would cause us to reject the hypothesis, $p \geqq \frac{1}{2}$, in favor of $p < \frac{1}{2}$. Here we have a one-tail test. The decision to use either a one-tail or a two-tail test is usually guided by the aim of the experimenter and particularly by the alternative to be accepted if the null hypothesis is rejected. The size of the type II error, β, depends on the acceptance region, whether it includes one-tail or no-tail. Other things being equal, we select the acceptance region that minimizes β for a given α.

(3) In Table 6–3 the probabilities of type II errors for different values of p are shown. This table shows, for example, that if the new formula really cuts down the frequency of after-effects from 0.50 to 0.40, the experimenter has a probability of 0.87 of not discovering it, and if the reduction was from 0.50 to 0.25 there still would be a chance of 0.38 of not finding it out. If α is to be kept at 0.02 then such large probabilities of a type II error should be

Table 6–3

True value of p	Probability of wrong decision
0.40	0.87
0.30	0.58
0.25	0.38
0.20	0.20
0.10	0.01

avoided if possible. The experimenter has three quantities at his disposal, α, β, and N, the number of observations (patients), to be taken. The choice of any two determines the third. It would be possible for him to select α and β and then determine N. If this were done it might happen that the required number of observations is larger than he is able to handle, and if this is so, the researcher could find this out before he wastes his time with fruitless observations.

(4) Tests of hypotheses are sometimes called *tests of significance*, and the probability, α, is termed the *significance level*. A probability value is said to be *significant* if it causes us to reject the null hypothesis. In many statistical experiments such as those discussed in Chapter 8, the type I error, α, or its equivalent, the significance level, (often expressed as a per cent) is selected in advance and then the rejection region is determined from it as either a one-tail or a two-tail region. Of course the choice of a P value for significance is for the experimental statistician to decide. Sometimes it is quite arbitrary; more often it is guided by weighty considerations. See Section 8.12 for further remarks on this topic.

6.13. The Binomial Distribution

The binomial theorem is not unfamiliar to students of high-school algebra. In its elementary form, it may be written as the equation:

$$(q + p)^N = q^N + \frac{N!}{1!(N-1)!} q^{N-1}p + \frac{N!}{2!(N-2)!} q^{N-2}p^2$$

$$+ \frac{N!}{3!(N-3)!} q^{N-3}p^3 + \cdots + \frac{N!}{x!(N-x)!} q^{N-x}p^x + \cdots + p^N, \tag{6.6}$$

where N is a positive integer. The $(x + 1)$th term, or general term of the expansion, is seen to be identical with Formula (6.5), which represents the probability for exactly x successes in N trials. Therefore, the terms of the binomial expansion in the right member of (6.6) represent, in order, the probabilities for exactly 0, 1, 2, 3, ..., N successes in N trials. If in Formula (6.6) we let $q = p = \frac{1}{2}$ and $N = 10$, we obtain:

$$\left(\frac{1}{2} + \frac{1}{2}\right)^{10} = \left(\frac{1}{2}\right)^{10} + \frac{10!}{1!9!} \left(\frac{1}{2}\right)^9\left(\frac{1}{2}\right) + \frac{10!}{2!8!} \left(\frac{1}{2}\right)^8\left(\frac{1}{2}\right)^2 + \cdots$$

$$+ \frac{10!}{9!1!} \left(\frac{1}{2}\right)\left(\frac{1}{2}\right)^9 + \left(\frac{1}{2}\right)^{10}$$

$$= \frac{1}{1024} + \frac{10}{1024} + \frac{45}{1024} + \cdots + \frac{10}{1024} + \frac{1}{1024}.$$

The 11 fractions above are those shown in Table 6–1, where they represented the probabilities for just 0, 1, 2, ..., 10 heads to appear when a penny is tossed 10 times.

We define a *binomial distribution* as one for which the frequencies are proportional to the successive terms of the binomial expansion (6.6). The name *Bernoulli distribution* is also employed, in honor of James Bernoulli (1654–1705), who first discovered it. The binomial distribution is theoretical, and may be used as a model for certain distributions arising in practice. A graphical representation of such a model appears in Figure 6–1.

6.14. The Mean of a Binomial Distribution

The theoretical mean μ, of a binomial distribution is easily derived with the aid of Formula (5.8).

$$\mu = E(x) = \sum_{i=1}^{n} p_i x_i. \tag{6.7}$$

In this formula p_i becomes the binomial probability

$$P(x) = \frac{N!}{x!(N-x)!} \, p^x q^{N-x}, \tag{6.8}$$

and the values x_i, become the values 0, 1, 2, ..., N. Thus

$$\mu = \sum_{x=0}^{N} P(x)x$$

$$= \sum_{x=0}^{N} \frac{N!}{x!(N-x)!} \, p^x q^{N-x} x. \tag{6.9}$$

When $x = 0$ the first term of this sum is zero, so that

$$\mu = \sum_{x=1}^{N} \frac{N!}{x!(N-x)!} \, p^x q^{N-x} x$$

$$= Np \sum_{x=1}^{N} \frac{(N-1)!}{(x-1)!(N-x)!} \, p^{x-1} q^{N-x}$$

$$= Np(q+p)^{N-1}.$$

But $p + q = 1$, hence

$$\mu = Np. \tag{6.10}$$

Thus if an unbiased coin is tossed 100 times the expected number of heads or theoretical mean is $100 \times \frac{1}{2}$ or 50. Although we really do not "expect" head to appear just 50 times, we do expect, in repeated experiments of tossing this coin 100 times, that the number of heads will average about 50.

6.15. The Standard Deviation of a Binomial Distribution

The variance of a discrete distribution is defined in Formula (5.9):

$$\sigma^2 = E(x - Np)^2 = \Sigma \, P(x)(x - Np)^2.$$

The products following the summation sign may be obtained in a convenient form for summing by writing

$$(x - Np)^2 = x^2 - 2xNp + N^2 p^2$$

$$= x + x(x-1) - 2Npx + N^2 p^2.$$

We have then four sums of products where $P(x)$ is given by (6.8):

(1) $\quad \sum_{x=0}^{N} P(x)x = Np,$ from the preceding section;

(2) $\quad \sum_{x=0}^{N} P(x)x(x-1) = N(N-1)p^2 \sum_{x=2}^{N} \frac{(N-2)!}{(x-2)!(N-x)!} p^{x-2}q^{N-x}$

$= N(N-1)p^2(q+p)^{N-2}$

$= N^2p^2 - Np^2;$

(3) $\quad \sum_{x=0}^{N} P(x)(-2Npx) = -2Np \sum_{x=0}^{N} P(x)x$

$= -2N^2p^2;$

(4) $\quad \sum_{x=0}^{N} P(x)N^2p^2 = N^2p^2 \sum_{x=0}^{N} P(x)$

$= N^2p^2(q+p)^{N}$

$= N^2p^2.$

When we add these four results we obtain

$$Np + N^2p^2 - Np^2 - 2N^2p^2 + N^2p^2 = Np - Np^2$$

$$= Np(1-p)$$

$$= Npq.$$

Thus the variance

$$\sigma^2 = Npq, \tag{6.11}$$

and the standard deviation

$$\sigma = \sqrt{Npq}. \tag{6.12}$$

If a penny is tossed 100 times, the theoretical standard deviation will be $\sqrt{100 \times \frac{1}{2} \times \frac{1}{2}}$, or 5. This means that when 100 pennies are tossed a large number of times, the average deviation of the number of heads from the expected value, 50, as measured by σ, will be 5.

6.16. Other Properties of the Binomial Distribution

The mean and the standard deviation of the distribution defined by $(q+p)^{N}$ are but two important measures associated with the binomial distribution. It may be shown that the most probable number of successes or the theoretical mode is the positive integer, x, determined by the double inequality:

$$Np - q \leq x \leq Np + p. \tag{6.13}$$

Since p and q are positive fractions having a sum unity, it follows that the most probable value, to within a proper fraction, is Np. In case $Np - q$ and $Np + p$ are integers, the equality signs hold, and there will be two adjacent modal values with equal probability.

As one illustration, consider the most probable number of black balls drawn when 10 drawings with replacements are made from a box containing 2 black and 4 white balls. Here $N = 10$, $p = \frac{1}{3}$, $q = \frac{2}{3}$; x is the integer determined by the inequality:

$$\frac{10}{3} - \frac{2}{3} \leq x \leq \frac{10}{3} + \frac{1}{3}.$$

Hence, $x = 3$.

As a second illustration, we find the most probable number, when 20 drawings are made, to be the integer x such that:

$$\frac{20}{3} - \frac{2}{3} \leq x \leq \frac{20}{3} + \frac{1}{3},$$

so that x may be either 6 or 7.

The third and fourth moments of this important distribution may also be found. (See Sections 4.18, 19.) They are defined as

$$\alpha_3 = \Sigma\, P(x)(x - Np)^3,$$

and
$$\alpha_4 = \Sigma\, P(x)(x - Np)^4.$$

By methods similar to the one used in the previous section one can prove that

$$\alpha_3 = \frac{q - p}{(Npq)^{\frac{1}{2}}}\,; \tag{6.14}$$

$$\alpha_4 = \frac{1}{Npq} - \frac{6}{N} + 3. \tag{6.15}$$

6.17. Binomial Probability Calculations

Except for fairly small N, say, $N \leq 10$, and simple probability values such as $p = 0.3$, $p = 0.5$, etc., the calculation of binomial probabilities can be wearisome if not practically impossible for the average student in statistics. An effective method of approximating them is discussed in Section 7.10.

On the other hand, the necessity for the calculation of binomial and cumulative binomial probabilities for large N and useful values of p has been largely eliminated by the publication of tables of these probabilities. The two most useful tables in the United States are *Tables of the Binomial Probability Distribution* published by the National Bureau of Standards (Reference 24) and the *Tables of the Cumulative Binomial Probability Distribution* published by the Harvard University Computation Laboratory (Reference 21). The former give both individual terms and cumulated terms for values of p ranging from 0.01 to 0.50 (by increments of 0.01) and for N ranging from 2 to 49. The latter give cumulated terms for similar values of p and for values of N up to 1000 (by varying increments). The individual terms can be derived from the cumulated terms by simple subtraction.

6.18. Binomial Probability Paper

Besides the tables mentioned above, attention is also called to the possibilities of the type of graph paper known as binomial probability paper. It is described in a paper by J. W. Tukey and F. Mosteller, "The Use and Usefulness of Binomial Probability Paper," in the *Journal of the American Statistical Association*, Vol. 44 (1949), p. 174.

EXERCISES

1. If a club has four candidates for president, three for vice-president, and two for secretary-treasurer, in how many ways may the officers be elected?

2. An examination contains three groups of questions. Group A contains five questions, Group B, two, and Group C two. One question is to be answered from each group. In how many different ways may a student select his questions?

3. Find the number of permutations that can be made from the letters a, b, c, d, e, and f taken (a) four at a time? (b) all at a time.

4. How many different arrangements can be made from the letters of the word *distance* by taking (a) five at a time? (b) all at a time?

5. Twenty sprinters compete in a race for which there are first, second, and third prizes. In how many ways may the prizes be awarded?

6. Find the number of permutations that can be made from the letters of the word *Oshkosh* taken all at a time.

7. When a penny was tossed eight times in succession, head appeared 3 times and tail 5 times in the following order: *THHTHTTT*. In how many other orders could they have appeared?

8. A true-false test consisted of ten statements of which seven were false and three were true. If a student knew this but merely guessed at the answers, in how many ways might he have listed his answers?

9. In how many ways may a committee of four be chosen from a club of nine members?

10. In an examination paper, any three questions may be omitted from the ten questions given. In how many ways may selections be made?

11. A witness to a bank robbery said that the license number of the criminals' automobile was a six-figure number of which the first three figures were 487. He did not recall the last three figures, but was positive that all three were different. How many automobile license numbers must the police check?

12. (a) In how many ways may an assignment of five problems be made from a group of twelve problems?
(b) How many times will the most difficult problem be assigned?

13. In how many ways may a committee consisting of five men and four women be chosen from ten men and seven women?

14. In how many ways may a party of five men be chosen from a company of nine men? In how many of these parties will a particular man, Mr. Smith, be included?

15. In a certain club there are twice as many men as women. If five committee members are chosen by lot, what is the probability that four are men and one is a woman?

16. If, in general, 30 per cent of patients afflicted with a certain disease die from it, what is the probability that just two die in a group of five?

17. Refer to Exercise 5.16. If eight New Englanders are selected at random, what is the probability that just three belong to blood group A? Compute your answer to three decimal places.

18. On an average, 10 per cent of the wooden rods used in a certain product are found to be too knotty for use. What is the probability that in a bundle of ten rods just five are too knotty?

19. If a thumb tack lands point-up 60 per cent of the time that it falls, what is the probability that when five thumb tacks fall (a) at least four land point-up? (b) not more than two will land point-up?

20. A coin suspected of bias is to be tossed ten times. If we choose $P < 0.10$ as the probability to be used for rejecting the hypothesis, $p = \frac{1}{2}$, and two heads appear, shall we conclude that the coin is biased?

21. We are to test the hypothesis that thumb tacks when dropped on the floor, land point-up, on an average, not more than 60 per cent of the time. If eight of them are dropped and six land point-up, shall we accept or reject this hypothesis? Let $\alpha = 0.05$. Note that $(0.6)^6 = 0.0467$ (approximately).

22. In cases of high blood pressure, a certain drug has been found to reduce it substantially in about 60 per cent of the cases. A modification of the drug is expected to increase the percentage of beneficial results and is tried experimentally on ten typical patients. If nine of these show substantial improvement and we choose $\alpha = 0.05$, shall we reject or accept the hypothesis $p = 0.60$? Use a one-tail test. Note that $(0.60)^9 = 0.0101$ (approximately).

23. Would it be possible to conclude that a penny is biased with just five tosses if $\alpha = 0.05$? Here $H_0: p = \frac{1}{2}$.

24. In Exercise 22 assume $H_0: p = 0.7$, and that the rejection values are 9 and 10 improvements out of 10 patients treated. If it happened that p really equalled 0.5, what is the probability, β, of a type II error? Note that $P(0) + P(1) + \cdots + P(8) = 1 - [P(9) + P(10)]$.

25. In general, 75 per cent of certain seeds germinate when planted properly. A sample of ten such seeds are subjected to very heavy air pressure before planting and it is found that nine of them germinate. Let $H_0: p = \frac{3}{4}$, $H_1: p > \frac{3}{4}$, and $\alpha = 0.05$. (a) Can we conclude that air pressure increases germination? Use a one-tail test. (b) If all seeds germinate could we make the same conclusion as in (a)? (c) What changes in the experiment would you suggest in view of your answers in (a) and (b)? Note that $(\frac{3}{4})^9 = 0.0751$.

26. Refer to the preceding exercise and alter it as follows: Suppose $H_0: p = 0.60$, $H_1: p > 0.60$, $N = 10$, and rejection of H_0 occurs when 9 or 10 seeds germinate. (a) What is the value of α? (b) If the true p were 0.80 what is the value of β? Note: $(0.8)^9 = 0.134$ approximately.

27. William Shanks (1812–1882) computed the number π to 707 decimal places. How many times would you expect him to have found the digit 0?

28. A penny is tossed 64 times. Find (a) the expected number of heads; (b) the theoretical standard deviation.

29. One hundred cuts are made of a pack of 52 playing cards. What is the expected number of spades appearing? What is the standard deviation?

30. Ten pennies are tossed. What is the probability for a deviation of less than two heads from the expected number?

31. A pair of dice is thrown 100 times. What is the expected number of times that the sum five appears? What is the standard deviation?

32. If a coin is tossed fifteen times what is the most probable number of heads?

33. Find the most probable number of sevens obtained when a pair of dice is thrown five times.

34. A pair of dice is tossed 50 times. (a) What is the mean number of times that the sum eight appears? (b) What is the expected value of the square of the deviation from the mean? (c) What is the most probable number of times that the sum eight appears?

35. Five dimes are tossed and the player is to receive all the dimes that show heads. What is his expected value?

36. From the Bureau of Standards *Table of the Binomial Probability Distribution* find the following:

(a) $P(27)$ when $N = 45$, $p = 0.43$;

(b) $P(4)$ when $N = 35$, $p = 0.17$;

(c) $\sum\limits_{x=0}^{7} P(x)$ when $N = 42$, $p = 0.30$;

(d) $\sum\limits_{x=3}^{10} P(x)$ when $N = 24$, $p = 0.25$.

37. From the Harvard *Table of the Cumulative Binomial Probability Distribution* find the following:

(a) $\sum\limits_{x=130}^{400} P(x)$ when $N = 400$, $p = 0.30$;

(b) $\sum\limits_{x=38}^{100} P(x)$ when $N = 100$, $p = 0.60$;

(c) $\sum\limits_{240}^{260} P(x)$ when $N = 500$, $p = 0.50$;

(d) $P(140)$ when $N = 440$, $p = 0.27$.

THE NORMAL
FREQUENCY FUNCTION

7

"You boil it in sawdust; you salt it in glue:
You condense it with locusts and tape:
Still keeping one principal object in view—
To preserve its symmetrical shape."
LEWIS CARROLL
The Hunting of the Snark

7.1. Introduction

In the preceding chapter we derived the important formula

$$P(x) = \frac{N!}{x!(N-x)!} p^x q^{N-x} \tag{7.1}$$

for the binomial frequency function. Given N and p, the probability, $P(x)$, for x successes can be found by substituting the value of x in this formula. Thus the probability or theoretical relative frequency, $P(x)$, depends upon the variable, x, or, as we say in mathematics, *P is a function of x.* In statistics we derive and use many frequency functions. These enable us to compute probabilities corresponding to given values or ranges of values of the variable, denoted by x, but often designated by some other letter such as t, u, y, and so forth. Sometimes (see Section 13.14) the frequency function involves two or more variables. Perhaps the most basic frequency function in the entire field of statistics is the one that forms the title of this chapter.

7.2. The Normal Frequency Function

From the preceding formula, (7.1) we find that the binomial frequency function yielding the probability of obtaining x heads when an unbiased coin is tossed 10 times is

$$P(x) = \frac{10!}{x!(10-x)!} (\tfrac{1}{2})^x (\tfrac{1}{2})^{N-x}. \tag{7.2}$$

The probability diagram corresponding to this formula is shown as Figure 6–1. We now reproduce this graph in a slightly different form by using artificial class boundaries $-\tfrac{1}{2}$ to $\tfrac{1}{2}$, $\tfrac{1}{2}$ to $1\tfrac{1}{2}$, $1\tfrac{1}{2}$ to $2\tfrac{1}{2}$, ..., $9\tfrac{1}{2}$ to $10\tfrac{1}{2}$, in place

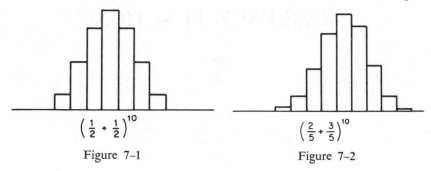

$$\left(\tfrac{1}{2}+\tfrac{1}{2}\right)^{10} \qquad\qquad \left(\tfrac{2}{5}+\tfrac{3}{5}\right)^{10}$$

Figure 7–1   Figure 7–2

of the class marks 0, 1, 2, ..., 10 respectively, and by using rectangles instead of vertical lines. (See Figure 7–1.) Since the base of each rectangle is of unit length, the area of each rectangle is numerically equal to its height. These changes give us a histogram that has the desirable property of permitting the areas of the rectangles, rather than the heights, to represent the probabilities.

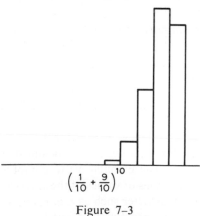

$$\left(\tfrac{1}{10}+\tfrac{9}{10}\right)^{10}$$

Figure 7–3

It will be noted that Figure 7–1 is symmetrical and somewhat bell-shaped. This symmetry is due to the fact that $p = q$. Figure 7–2 represents the probability diagram for $N = 10$ as before, but with $p = \tfrac{3}{5}$, $q = \tfrac{2}{5}$. Note that the figure is very nearly symmetrical and bell-shaped. In Figures 7–3, 4, and 5, $p = \tfrac{9}{10}$ and $q = \tfrac{1}{10}$, but N is 10 for the first named figure, 20 for the second, and 50 for the third. The extremely asymmetrical histogram of Figure 7–3 becomes more symmetrical and bell-shaped in Figures 7–4 and 5, although the last two named have long tails extending to the left. There the heights of the rectangles are so small as not to be reproducible in the text.

If we refer to Figure 7–10 where $N = 400$, $p = q = \frac{1}{2}$, we have clearly suggested a symmetrical bell-shaped curve with extremely long tails (not shown) to left and right that cover the intervals, 0 to 170 heads and 230 to 400 heads, all corresponding to very unlikely outcomes. These facts suggest

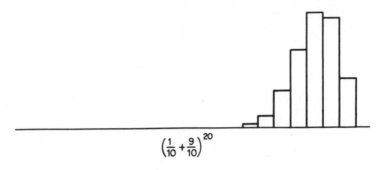

$$\left(\tfrac{1}{10} + \tfrac{9}{10}\right)^{20}$$

Figure 7–4

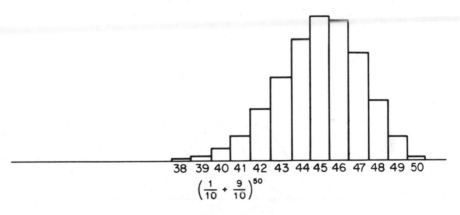

38 39 40 41 42 43 44 45 46 47 48 49 50

$$\left(\tfrac{1}{10} + \tfrac{9}{10}\right)^{50}$$

Figure 7–5

that as N increases indefinitely the binomial histogram approaches as a limit the area under a certain bell-shaped curve and that this approach becomes more rapid the more nearly equal p and q are. In fact it is possible to prove mathematically a very important theorem, but the proof lies beyond the scope of this book. We recall first that for a binomial distribution (Sections 6.14, 15) the mean, $\mu = Np$ and the standard deviation, $\sigma = \sqrt{Npq}$. The theorem is the following:

A binomial frequency function

$$P(x) = \frac{N!}{x!\,(N-x)!}\, p^x q^{N-x},$$

in which N becomes infinitely large, approaches as a limit the so-called normal frequency function,

$$f(x) = \frac{1}{\sqrt{2\pi Npq}} \, e^{-\frac{(x-Np)^2}{2Npq}}$$

or

$$f(x) = \frac{1}{\sqrt{2\pi}\,\sigma} \, e^{-\frac{1}{2}\left(\frac{x-\mu}{\sigma}\right)^2}. \tag{7.3}$$

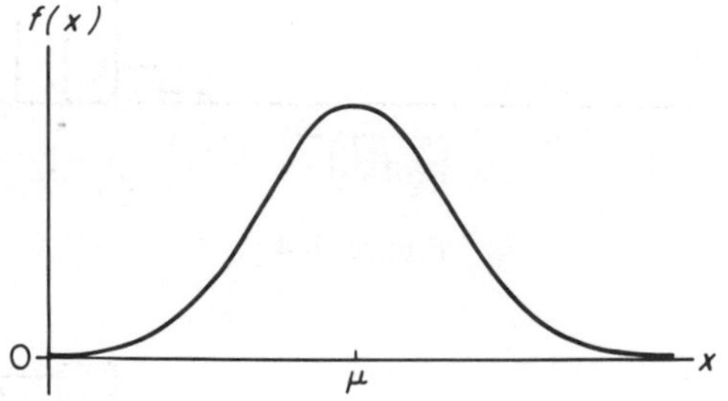

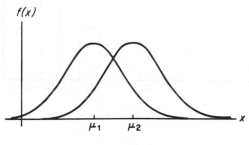

Figure 7–6

Figure 7–7

    The graph of the normal frequency function is shown in Figure 7–6. It is called the *normal curve.*

    In Formula (7.3) π is the familiar constant whose approximate value is 3.14159; e stands for another important constant, the base of the natural system of logarithms whose approximate value is 2.71828. In addition to these constants there are two *parameters,* μ and σ, which determine the position and relative proportions of the normal curve. Thus if two populations defined by normal frequency functions had different means, μ_1 and μ_2, but identical standard deviations, $\sigma_1 = \sigma_2$, their graphs would appear like those in Figure 7–7. If the populations had identical means, $\mu_1 = \mu_2$, but different standard deviations, σ_1 and σ_2, their graphs would appear as shown in Figure 7–8. If both the means and the standard deviations were different we would have curves like those in Figure 7–9. In all three diagrams the areas under the curves are equal to one unit. In Figures 7–8 and 9 there is a

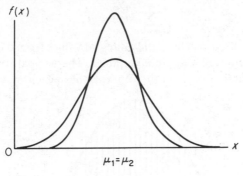

Figure 7–8

difference in the spread of the curves due to the difference in the σ's. Thus a normal distribution is completely characterized by the two parameters, μ and σ. These parameters bear to a normal population the same relation that the statistics $\bar{x}$ and s bear to a sample drawn from such a population.

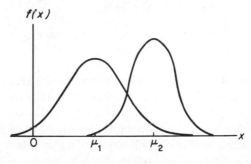

Figure 7–9

The normal frequency function is used as a model by means of which many frequency distributions arising in practical work may be described and analyzed. Thus the graph of Figure 3–1 suggests that the sample of 462 head lengths might reasonably be assumed to have been drawn from an infinite population of head lengths, the distribution of which is governed by a normal frequency function. It might be remarked now that various tests exist for determining the validity of such an assumption. (See Section 7.13.)

The normal curve extends to infinity in both directions and in so doing approaches infinitely close to the x-axis without ever actually touching it. This property is described by stating that the curve is *asymptotic* to the x-axis.

7.3. Area and Probability

The total area of a binomial histogram equals unity since the area of each rectangle represents a probability and the sum of these areas,

$$\sum_{x=0}^{N} P(x) = 1.$$

We note that because x can have only integral values from 0 to N, the binomial function is a function of a discrete variable. The normal frequency function, however, is a function of a continuous variable which may assume *any* real value.

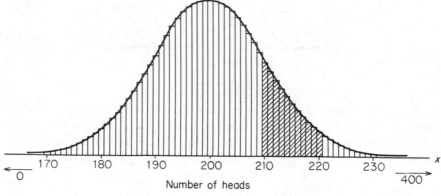

Fig. 7–10.   Binomial histogram and normal curve corresponding to 400 tosses of a penny.

By the methods of calculus it can be shown that the area under the normal curve also equals unity. Because of this fact any partial area under the curve is interpreted as a probability. In order to clarify this last statement let us refer to Figure 7–10 where the probability, for example, of getting anywhere from 210 to 220 heads inclusive in 400 tosses of a true coin is given by the sum of the areas of the rectangles from 210 to 220 inclusive. Since N is a large number, the form of the histogram is nearly that of a normal curve so that we should obtain an excellent approximation to the desired probability by taking the area under the corresponding normal curve from 210 to 220. Since $N = 400$, $p = q = \frac{1}{2}$, it follows that $\mu = 400 \times \frac{1}{2} = 200$, $\sigma = \sqrt{400 \times \frac{1}{2} \times \frac{1}{2}} = 10$, and the equation of the approximating normal curve (7.3) becomes

$$f(x) = \frac{1}{\sqrt{2\pi}\,10}\, e^{-\frac{1}{2}\left(\frac{x-200}{10}\right)^2} \tag{7.4}$$

The problem of finding the exact area under any portion of the normal curve is readily solved by calculus methods but in order to eliminate unnecessary calculations, tables exist which enable one to find readily any desired area under the normal curve. The use of these tables is discussed in Section 7.6. The task of calculating directly from the binomial formula (7.1) the probability of 210 to 220 heads inclusive would be very laborious.[1]

$$P(210) + P(211) + \cdots + P(220) = \frac{400!}{210!\,190!}\,(\tfrac{1}{2})^{210}(\tfrac{1}{2})^{190}$$

$$+ \frac{400!}{211!\,189!}\,(\tfrac{1}{2})^{211}(\tfrac{1}{2})^{189} + \cdots + \frac{400!}{220!\,180!}\,(\tfrac{1}{2})^{220}(\tfrac{1}{2})^{180}.$$

(7.5)

7.4. The Standard Form

The parameters μ and σ, characterize a given normal distribution as we have already noted. Thus if the statures of adult American males were

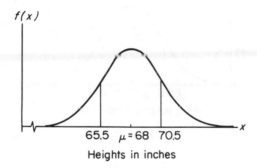

Figure 7–11

distributed normally with a mean, let us say, of 68 inches and a standard deviation of 2.5 inches, this population of statures would be characterized by the frequency function

$$f(x) = \frac{1}{\sqrt{2\pi}\,2.5}\,e^{-\frac{1}{2}\left(\frac{x-68}{2.5}\right)^2}$$

(7.6)

and represented graphically by Figure 7–11. In this figure the unit of measure is one inch. The distance between the mean, 68, and the two statures 65.5 and 70.5 inches is 2.5 inches in each case—the value of σ.

If the population of English head lengths, assumed to be normal, had $\mu = 192$ millimeters and $\sigma = 6.5$ millimeters, its frequency function would be

$$f(x) = \frac{1}{\sqrt{2\pi}\,6.5}\,e^{-\frac{1}{2}\left(\frac{x-192}{6.5}\right)^2},$$

(7.7)

[1] If one has available *Tables of the Cumulative Binomial Probability Distribution* (see Reference 21) one can quickly find the above sum to be 0.15094.

and its graph, that shown in Figure 7–12. Here the unit of measure is one millimeter.

In order to make use of probability tables it is desirable to employ a frequency function that is independent of the units used. To this end, as in Section 4.16, we introduce the *standard variable* or *standard deviate*, z, where

$$z = \frac{x - \mu}{\sigma}.$$ (7.8)

z yields the number of σ's by which x deviates from μ. Since for a given population $x - \mu$ and σ are measured in the same units, their quotient, z, is a pure number. As an example, let the head length, x, of an English criminal

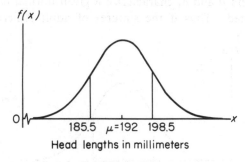

Head lengths in millimeters

Figure 7–12

be 205 mm. Then $x - \mu = 205 - 192 = 13$ mm; $\sigma = 6.5$ mm; hence $z = 13/6.5 = 2$, a pure number. When the standard variable is used in place of x, the effect, graphically, is to place the origin (zero point) of z at the mean, μ, of x and to use σ as a horizontal unit of measure. It can be proved that the area under this transformed curve remains equal to unity. The function, $f(x)$, defined in (7.3) now is transformed into the *standard form* of the normal frequency function,

$$\phi(z) = \frac{1}{\sqrt{2\pi}} e^{-z^2/2}.$$ (7.9)

The relations just described between the curves of $f(x)$ and $\phi(z)$ are illustrated in Figure 7–13. The two points, $\mu_x + \sigma$ and $\mu_x - \sigma$, each at a σ's distance from μ_x in the upper graph correspond to the two points $+1$ and -1 each at a unit distance from the mean, $\mu_z = 0$. It can be proved easily by the methods of the calculus that the two *points of inflection* on the curve, that is, points at which the curve changes from one that is concave downward to one that is concave upward, are exactly at a distance of σ_x from μ_x (upper curve) and at a distance $\sigma_z = 1$ from $\mu_z = 0$ (lower curve).

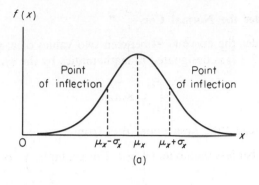

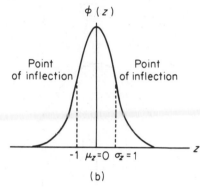

Figure 7–13

7.5. Ordinates of the Normal Curve

The values of the ordinate, $\phi(z)$, the height of the curve, corresponding to the variable z, are tabulated for values of z ranging numerically from $z = 0$ to $z = 4.00$ in Table C of the Supplementary Tables.

Thus,   for $z = 0$, $\phi(z) = 0.3989$,

         for $z = 1.00$, $\phi(z) = 0.2420$,

         for $z = 2.78$, $\phi(z) = 0.0084$,

and   for $z = -0.84$, $\phi(z) = 0.2803$.

Here, because of symmetry $\phi(z) = \phi(-z)$. By making use of such values, one can plot a very accurate graph of the standard normal curve. (See Exercise 7–11.)

Sometimes it is desirable to fit a normal curve to a frequency distribution arising from given data. In such a case for one method of fitting, a table of ordinates is necessary. (See Section 7.11.)

7.6. Areas Under the Normal Curve

The area under the curve (7.9) between two values of z, say $z = z_1$, and $z = z_2$, (Figure 7–14) is designated in mathematics by the symbol:

$$\int_{z_1}^{z_2} \phi(z)\, dz,$$

which may be read "the area under $\phi(z)$ from z_1 to z_2."[2] This area will always be a number less than one. (Why?) For example, $\int_0^{z_1} \phi(z)\, dz$ represents

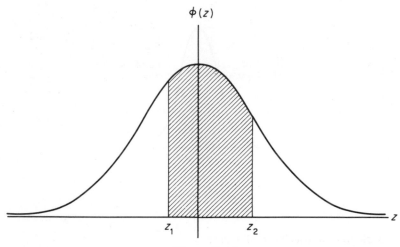

Figure 7–14

the area from the origin to z_1, $\int_{-2}^{3} \phi(z)\, dz$ represents the area from $z = -2$ to $z = +3$, and $\int_{1.50}^{\infty} \phi(z)\, dz$ represents the area from $z = 1.50$ to infinity, that is, it represents the entire area to the right of 1.50.

Table D of the Supplementary Tables gives values of the areas under the standard normal curve to the right of any non-negative value of z, say z_1, that is, it gives values of $\int_{z_1}^{\infty} \phi(z)\, dz$. Because of the symmetry of the curve with respect to the vertical or $\phi(z)$-axis, the area from any z, say z_1, to $+\infty$, equals the area from $-\infty$ to $-z_1$. (Figure 7–15.) Thus $\int_{z_1}^{\infty} \phi(z)\, dz = \int_{-\infty}^{-z_1} \phi(z)\, dz$.

[2] More precisely, this symbol, borrowed from calculus, is read "the definite integral of $\phi(z)\, dz$ from z_1 to z_2."

Note that we always express the interval for the area, from the left end-point to the right end-point. If we do this, the area under $\phi(z)$ can be shown always to be a positive number. From Table D one can easily verify that

$$\int_{0.50}^{\infty} \phi(z)\,dz = 0.3085, \qquad \int_{2.17}^{\infty} \phi(z)\,dz = 0.0150,$$

and so on.

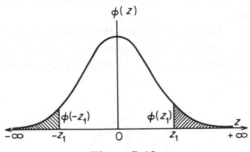

Figure 7–15

All areas under the curve other than those given in the table can be easily derived from them. Thus,

$$\int_{1.50}^{2.00} \phi(z)\,dz = \int_{1.50}^{\infty} \phi(z)\,dz - \int_{2.00}^{\infty} \phi(z)\,dz$$

$$= 0.0668 - 0.0228 = 0.0440.$$

$$\int_{-1.50}^{\infty} \phi(z)\,dz = 1 - \int_{-\infty}^{-1.50} \phi(z)\,dz$$

$$= 1 - \int_{1.50}^{\infty} \phi(z)\,dz = 1 - 0.0668 = 0.9332.$$

$$\int_{-1.32}^{2.13} \phi(z)\,dz = 1 - \left[\int_{-\infty}^{-1.32} \phi(z)\,dz + \int_{2.13}^{\infty} \phi(z)\,dz \right]$$

$$= 1 - \left[\int_{1.32}^{\infty} \phi(z)\,dz + \int_{2.13}^{\infty} \phi(z)\,dz \right]$$

$$= 1 - (0.0934 + 0.0166) = 0.8900.$$

The total area under the curve may be expressed as:

$$\int_{-\infty}^{\infty} \phi(z)\,dz = 2 \int_{0}^{\infty} \phi(z)\,dz = 1.$$

The symbol for *infinity*, ∞, indicates that the limit has become infinitely great.

7.7. Properties of the Standard Curve

Let us set down again for convenient reference the equation of the normal frequency curve in standard form:

$$\phi(z) = \frac{1}{\sqrt{2\pi}} e^{-z^2/2}. \tag{7.10}$$

The following properties are of chief importance. Some of them have already been noted.

(1) *Symmetry.* The curve is symmetrical with respect to the $\phi(z)$-axis. This can be shown directly from Equation (7.10). If z were replaced by $-z$, $\phi(z)$ would remain unchanged. In other words, the ordinate $\phi(z)$ is the

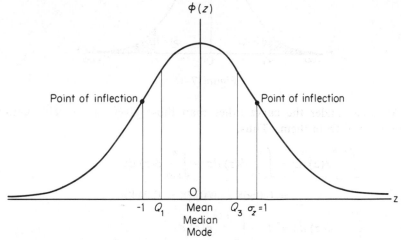

Fig. 7–16.   Important values associated with the normal frequency curve in standard form.

same at equal distances on either side of the origin. Statistically, this means that the arithmetic mean and the median of a normal frequency distribution coincide at the center of it. $\mu_z = 0$ corresponds to $x = \mu_x$.

(2) *Shape.* The exponent of e in $\phi(z)$ is negative, $-\frac{1}{2} z^2$. Hence, $\phi(z)$ is a maximum when $z = 0$; all other values of z make $\phi(z)$ smaller, since $e^{-z^2/2} = \dfrac{1}{e^{z^2/2}}$. The maximum value of $\phi(z)$ is, therefore,

$$\phi(0) = \frac{1}{\sqrt{2\pi}} = 0.3989.$$

As z increases numerically, $e^{-z^2/2}$ decreases; and when z becomes infinite, $\phi(z)$ approaches zero. Thus, the curve is asymptotic to the z-axis in both the positive and negative directions.

The *points of inflection* of the curve are the points at which the curve changes from concave downward to concave upward. By the methods of calculus it can be shown that the points of inflection are situated at a unit's distance from the $\phi(z)$-axis. This distance can be shown to be equal to the standard deviation of z, so that $\sigma_z = 1$. (Figure 7–16.)

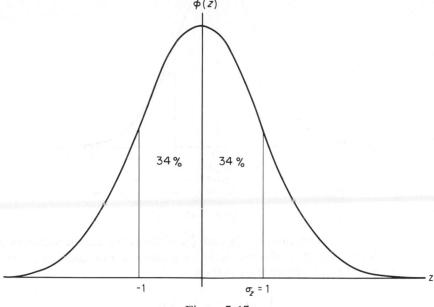

Figure 7–17

It is clear, then, from the preceding paragraphs that the standard curve has its maximum value $1/\sqrt{2\pi}$ units above the origin, that it is concave downward until $z = \pm 1$, when it becomes concave upward, and that it rapidly approaches, but never quite reaches, the z-axis. These properties determine its bell-shaped form.

(3) *Areas.* The total area under the curve has already been shown to be exactly one.

The area under the curve from $z = 1$ to $z = \infty$ is 0.1587, so that the area comprised within the interval -1 to $+1$,

$$\int_{-1}^{1} \phi(z)\, dz = 1 - 2(0.1587) = 0.6826.$$

Statistically, this means that about 68 per cent of normal variates deviate from their mean by less than one standard deviation. (Figure 7–17.) Similarly

$$\int_{-2}^{2} \phi(z)\, dz = 0.9544; \quad \text{and} \quad \int_{-3}^{3} \phi(z)\, dz = 0.9974.$$

The preceding values show that although the curve extends indefinitely to the left and to the right, it approaches the z-axis so closely that over 95 per cent of the area is included between the limits -2 and $+2$, and over 99.7 per cent of the area is included between -3 and $+3$. (Figure 7–18.)

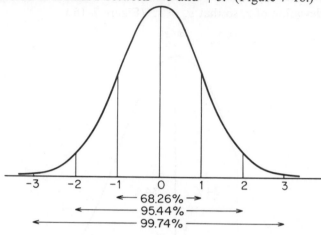

Figure 7–18

(4) *Quartiles.* The quartiles Q_1 and Q_3 of the curve are the values of z whose ordinates together with the $\phi(z)$-axis divide the area under the curve into four equal areas. From the equation

$$\int_{z_1}^{\infty} \phi(z)\, dz = 0.2500$$

we can, by inverse interpolation in Table D, find the value of z_1 corresponding to one-fourth the area. This value of z is found to be 0.6745, and represents Q_3. Hence, $Q_1 = -0.6745$ and $Q_3 = +0.6745$. (Figure 7–19.)

The semi-interquartile range or quartile deviation is obviously 0.6745. Fifty per cent of the area lies between -0.6745 and $+0.6745$.

7.8. $\phi(z)$ and $f(x)$

We have seen that the unit of measure along the x-axis is σ_x times as great as that along the z-axis. In other words, σ_x itself is the unit when we employ the variable z. That is why $\sigma_z = 1$. This is a very useful device and enables us to express easily many of the properties just enumerated, in terms of the original variable, x. For example, abscissas of the points of inflection of the curve (7.3) are $\mu \pm \sigma_x$. Therefore, 68 per cent of the variates have values lying between the limits $\mu + \sigma_x$ and $\mu - \sigma_x$. Less than 0.3 per cent of the variates deviate from the mean by more than $3\sigma_x$. Fifty per cent of the variates lie within the limits $\mu \pm 0.6745\sigma_x$.

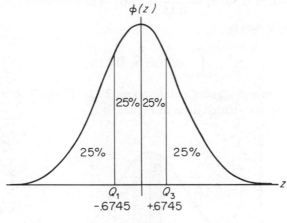

Figure 7–19

7.9. Some Applications

EXAMPLE 1.   Referring to the data for head lengths, Table 4–1, let us assume that head lengths are normally distributed and find the probability that a criminal chosen at random has a head length between 190.0 and 195.0 millimeters. Let $x_1 = 190.0$ and $x_2 = 195.0$. We have found that $\bar{x} = 191.8$ and $s = 6.48$. Since the sample is large we assume $\bar{x} = \mu$ and $s = \sigma$. Hence, by virtue of (7.8)

$$z_1 = \frac{x_1 - \bar{x}}{s_x} \qquad\qquad z_2 = \frac{x_2 - \bar{x}}{s_x}$$

$$= \frac{190.0 - 191.8}{6.48} \qquad\qquad = \frac{195.0 - 191.8}{6.48}$$

$$= -0.28 \qquad\qquad\qquad = 0.49.$$

Figure 7–20 shows the relation between the x- and z-curves.

$$\int_{-0.28}^{0.49} \phi(z)\, dz = 0.2982;$$

hence the probability sought is about 0.30.

EXAMPLE 2.   In the manufacture of washers to be used in radio receivers it has been found that the mean thickness, $\mu = 2.20$ mm and $\sigma = 0.15$ mm. All washers exceeding a thickness of 2.50 mm are rejected. What percentage can be expected to be discarded?

Since $x_1 - \mu = 0.30$, $z_1 = \dfrac{0.30}{0.15} = 2.00$. The probability for deviations greater than 0.30 mm is:

$$\int_{2.00}^{\infty} \phi(z)\, dz = 0.0228.$$

The percentage to be discarded will be about 2.28. The probability just found is represented by the shaded area in Figure 7–21.

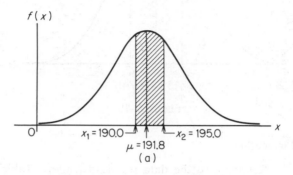

(a)

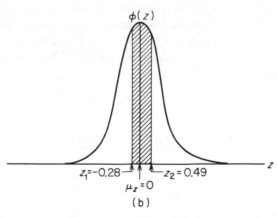

(b)

Figure 7–20

EXAMPLE 3. Assume that the scores in a Graduate Record Examination are normally distributed with $\mu = 500$ and $\sigma = 100$. Of 674 persons taking this examination it is desired to pass 550 of them. What should be the lowest score permitted for passing?

Solution: $\dfrac{550}{674} = 0.8160$ so that 81.60 per cent are to pass. The fraction failing is represented by the left-tail area shown in Figure 7–22(a). z_1, corresponding to the desired cut-off value, x_1, is a negative number

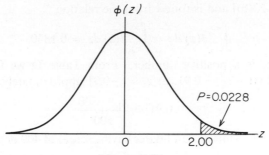

Figure 7–21

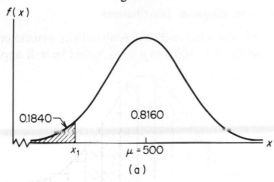

(a)

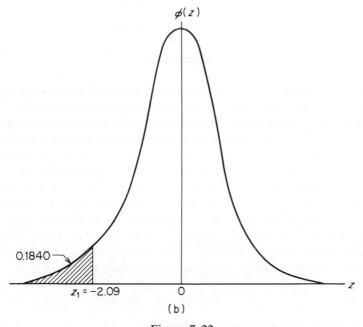

(b)

Figure 7–22

[See Figure 7–22(b)] and is found from the relation

$$\int_{-\infty}^{z_1} \phi(z)\, dz = \int_{-z_1}^{\infty} \phi(z)\, dz = 0.1840.$$

Note that $-z_1$ is a positive number. From Table D we find by inverse interpolation that $-z_1 = 0.91$, or $z_1 = -0.91$ approximately. Then

$$-0.91 = \frac{x_1 - 500}{100}$$

whence $x_1 = 409$. Thus all candidates having scores of 409 or above will pass.

7.10. Applications to Binomial Distributions

In Section 7.3 it was observed that probabilities associated with the binomial distribution for $N = 400$ and $p = \frac{1}{2}$, could be well approximated by

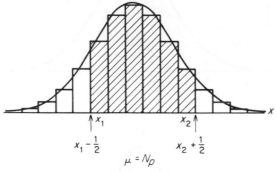

Figure 7–23

means of the normal distribution. In general when N is moderately large and p (or q) $\leq \frac{1}{2}$, but Np (or Nq) > 5, the normal approximation is valid. In this connection it is important to bear in mind that binomial frequencies are functions of a *discrete* variable whereas normal frequencies are functions of a *continuous* variable. For this reason a *correction for continuity* should be used.

Figure 7–23 represents a binomial histogram. Suppose that we seek the probability that the number of successes x, lies between x_1 and x_2 inclusive. Since each point representing a number of successes is the midpoint of the base of a rectangle, this probability is given by the sum of the areas of the rectangles within the interval from $x_1 - \frac{1}{2}$ to $x_2 + \frac{1}{2}$. The number $\frac{1}{2}$ is the *correction factor* for continuity. Then the sum of the probabilities for a number of successes from x_1 to x_2 inclusive is given by the integral

$$\int_{x_1 - \frac{1}{2}}^{x_2 + \frac{1}{2}} f(x)\, dx = \int_{z_1}^{z_2} \phi(z)\, dz$$

where z_1 and z_2 correspond to $x_1 - \frac{1}{2}$ and $x_2 + \frac{1}{2}$.

Thus $z_1 = \dfrac{x_1 - \frac{1}{2} - \mu}{\sigma}$ and $z_2 = \dfrac{x_2 + \frac{1}{2} - \mu}{\sigma}$ where $\mu = Np$ and $\sigma = \sqrt{Npq}$.

EXAMPLE 1. Let us calculate the sum of the probabilities $P(210) + P(211) + \cdots + P(220)$ of (7.5) by the method given above. $\mu = 200$, $\sigma = 10$, $x_1 = 210$, and $x_2 = 220$. Then

$$z_1 = \frac{210 - \frac{1}{2} - 200}{10} = 0.95, \qquad z_2 = \frac{220 + \frac{1}{2} - 200}{10} = 2.05.$$

$$\int_{0.95}^{2.05} \phi(z)\, dz = 0.1711 - 0.0202 = 0.1509.$$

The probability that in 400 tosses there would be a number of heads from 210 to 220 inclusive, is 0.1509. This agrees with the value given in the

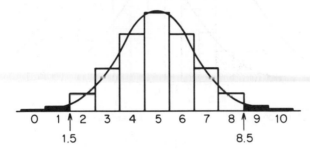

Figure 7–24

footnote accompanying Formula (7.5). The excellent agreement of the normal approximation (to four decimal places) with the exact binomial sum is due to the facts that N is a large number and $p = q$.

EXAMPLE 2. In Section 6.10 we calculated the probability for the number of heads in 10 tosses of a true coin to be 0, 1, 9, or 10. There we used the binomial probability formula to find

$$P(0) + P(1) + P(9) + P(10) = 0.0216.$$

Using the normal approximation (Figure 7–24) we find that $\mu = 5$, $\sigma = \sqrt{10 \times \frac{1}{2} \times \frac{1}{2}} = 1.58$. Let $x_1 = 1$ and $x_2 = 9$. Then

$$z_1 = \frac{1.5 - 5}{1.58} = -2.22, \qquad z_2 = \frac{8.5 - 5}{1.58} = 2.22.$$

Hence

$$P(x \leq 1) + P(x \geq 9) = \int_{-\infty}^{-2.22} \phi(z)\, dz + \int_{2.22}^{\infty} \phi(z)\, dz$$

$$= 2 \int_{2.22}^{\infty} \phi(z)\, dz = 0.0264.$$

The discrepancy between these two values, 0.0216 and 0.0264, is not important but is due to the small value of N and the use of the tail areas only.

EXAMPLE 3.   In the manufacture of certain experimental light bulb filaments, 20 per cent of them have been, as a rule, defective. A modification of the shape of the filament is to be tried out on a sample of 80 and if 10 or fewer are found defective, the hypothesis $H_0:p_0 = 0.20$ will be rejected in favor of $H_1:p_1 < 0.20$. On the other hand, if 11 or more are found to be

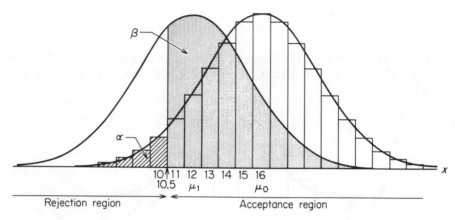

Figure 7–25

defective H_0 will be accepted, that is, the modification will be deemed to be no improvement. (a) What is the value of α? (b) If later extensive experience with the new filament showed that $p = 0.15$, what would be the size of β?

(a) Under H_0, $\mu_0 = 80 \times 0.20 = 16$, $\sigma_0 = \sqrt{80 \times 0.20 \times 0.80} = 3.58$;

$$z_1 = \frac{10.5 - 16}{3.58} = -1.54,$$

hence $P(x \leq 10) = P(z < -1.54) = 0.0618 = \alpha.$

(b) Under H_1, $\mu_1 = 80 \times 0.15 = 12$, $\sigma_1 = \sqrt{80 \times 0.15 \times 0.85} = 3.19$;

$$z_1 = \frac{10.5 - 12}{3.19} = -0.470;$$

$$P(x \geq 11) = P(z > -0.47) = 0.6808 = \beta.$$

Thus there is a 6 per cent chance of rejecting the hypothesis H_0, and deciding that the modification is better when it is not. On the other hand, there is a 68 per cent chance of accepting H_0 and not adopting the modified filament, when actually it is an improvement. (See Figure 7–25.)

EXAMPLE 4. In the United States 45 per cent of the population belong to blood group O. We are to test the hypothesis $H_0 : p_0 = 0.45$ against the alternative $H_1 : p_1 < 0.45$, on a sample of 200 persons of race R. If α is to be equal to 0.05, what is the critical number for the rejection of H_0? (b) What hypothesis should we accept if 80 such persons in the sample are found to belong to blood group O?

Here $\mu_0 = 200 \times 0.45 = 90$ and $\sigma_0 = \sqrt{200 \times 0.45 \times 0.55} = 7.38$. Then for a left-tail area of 0.05, $z = -1.645$. Since

$$-1.645 = \frac{x + \frac{1}{2} - 90}{7.38},$$

$x = 77.4$, hence the critical number (integer) is 77.4 (say 77). Therefore, if the number in the sample of 200 equals or is less than 77, we reject H_0, otherwise we accept it. Since 80 exceeds 77 we accept H_0.

7.11. Normal Curve Fitting

It is sometimes useful to fit a normal curve to a histogram constructed from a sample that appears to have arisen from a population that is reasonably normal. We assume that the population mean, μ, is very nearly equal to the

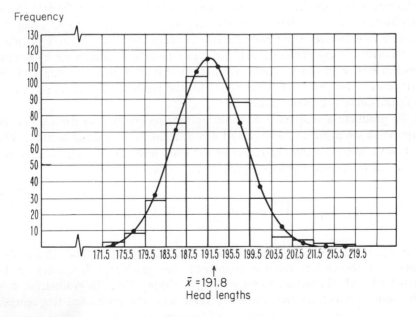

Fig. 7–26. Histogram and normal frequency curve for head lengths.

mean, $\bar{x}$, of the sample, and that the population standard deviation, σ, is very nearly equal to the corresponding statistic, s, of the sample. These assumptions seem to be valid in view of the fact that in order to have a histogram possible, the sample size, N, must be large.

The successive midpoints of the bases of the rectangles are labeled $x_1, x_2, ..., x_n$, and the latter are converted into standard variates, $z_1, z_2, ..., z_n$. From Table C we can find the ordinates $\phi(z_1), \phi(z_2), ..., \phi(z_n)$ of these midpoints. These represent the heights of the normal curve in standard form, and must be multiplied by the scale factor, NK/σ, in order to be plotted on the same scale as the histogram. Here k denotes the width of the class interval. It is good practice to plot also the maximum ordinate, obtained as $(Nk/\sigma)\phi(0)$.

Figure 7–26 shows the normal curve corresponding to the distribution of head lengths.

7.12. Graduation of Normal Data

If a normal curve fits a frequency distribution well, we may feel that the curve gives a better picture of what such a distribution would be "on the average" than the particular distribution itself does. We recall that a given distribution is conceived to be merely one sample chosen at random from infinitely many possible samples, and that these samples will exhibit variations or fluctuations from the ideal distribution. Conclusions drawn from the ideal distribution should be better "in the long run" than those based upon the single sample alone. Assuming the last statement to be true, we may calculate the theoretical frequencies corresponding to the actual frequencies by finding the area under the curve corresponding to that of a given rectangle of the histogram. This process of calculation for the purpose of *smoothing* the data to fit the curve is called *graduation*.

To graduate a frequency distribution that seems to stem from a normal population we again use $\mu = \bar{x}$ and $\sigma = s$, as in the preceding section. The successive values of the *end-points* of the class intervals, $x_1, x_2, ..., x_{n+1}$, are converted to standard values, $z_1, z_2, ..., z_{n+1}$. Let

$$p_i = \int_{z_i}^{z_{i+1}} \phi(z)\,dz, \qquad i = 1, 2, ..., n,$$

so that p_i is the probability for z to be between z_i and z_{i+1}. Since this probability is a theoretical relative frequency, the graduated frequency to be compared with the actual frequency, f_i, is simply Np_i. In graduation it is advisable to construct the frequency table with the graduated frequencies, $Np_1, Np_2, ..., Np_n$, matching the actual frequencies, $f_1, f_2, ..., f_n$, for purposes of comparison. (Table 7–1.)

Table 7–1

COMPUTATION FOR THE GRADUATION OF HEAD LENGTHS BY
MEANS OF THE NORMAL CURVE

1	2	3	4	5
End-value x_i	Actual frequency f_i	$\dfrac{x_i - \bar{x}}{s}$ z_i	$p_i = \displaystyle\int_{z_i}^{z_i+1} \phi(z)\, dz$	Graduated frequency Np_i
171.5		−3.13		
	3		.0051	2.4
175.5		−2.51		
	9		.0227	10.5
179.5		−1.90		
	29		.0716	33.1
183.5		−1.28		
	76		.1543	71.3
187.5		−.66		
	104		.2255	104.2
191.5		−.05		
	110		.2356	108.8
195.5		.57		
	88		.1673	77.3
199.5		1.19		
	30		.0811	37.5
203.5		1.80		
	6		.0281	13.0
207.5		2.42		
	4		.0066	3.0
211.5		3.04		
	2		.0011	.5
215.5		3.66		
	1		.0001	.0
219.5		4.27		
Total = 462		$\bar{x} = 191.8 \qquad s = 6.48$		Total = 461.6

7.13. Tests of Normality

Whether a given frequency distribution is normal or not is commonly decided by visual estimation—the frequency histogram looks fairly symmetrical and is bell-shaped, or the points plotted on arithmetic probability paper seem to lie along a straight line. (See Section 7.15.) It is possible to reinforce these qualitative procedures, or to supplant them, by tests which yield numerical measures.

The third moment, a_3, known as the skewness, and the fourth moment, a_4, of a sample have been defined in Sections 4.18 and 19. The mean deviation, $M.D.$, was defined in Section 2.9. When the standard deviate is employed, we define the mean deviation thus:

$$a = \frac{1}{N} \sum_{i=1}^{N} |z_i|. \qquad (7.11)$$

It can be demonstrated mathematically that for a standard normal distribution, the parameters corresponding to the above statistics have the following values:

$$\alpha_3 = 0;$$

$$\alpha_4 = 3;$$

$$\alpha = \sqrt{(2/\pi)} \text{ exactly;}$$

$$= 0.798 \text{ approximately.}$$

The last value is the basis for the relation given in Formula (4.13).

The discrepancies between a_3 and α_3, a_4 and α_4, and a and α are used as tests for the normality of a given sample. Tables of probability points for a_3, a_4, and a are available, and the methods of Chapter 8 used. (See E. S. Pearson and R. C. Geary; *Tests of Normality*, Biometrika Office, University College, London.)

The quantity *Chi-square* enables one to test for the "goodness of fit" of a normal curve to a given sample conjectured to be normal. (See Section 11.6.)

7.14. The Cumulative Normal Curve

In Section 3.8 we plotted the cumulative frequency diagram (Figure 3–10), corresponding to the frequency polygon (Figure 3–1) of head lengths. These

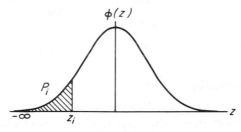

Figure 7–27

graphs are characteristic of large samples of normal type. In like manner there exists a *cumulative* normal curve associated with a normal curve. The former is plotted on this wise. The cumulative probabilities, $P_1, P_2, ..., P_n$, corresponding to the standardized end-values, $z_1, z_2, ..., z_n$, are obtained from Table D. (See Figure 7–27.)

$$P_i = \int_{-\infty}^{z_i} \phi(z)\, dz.$$

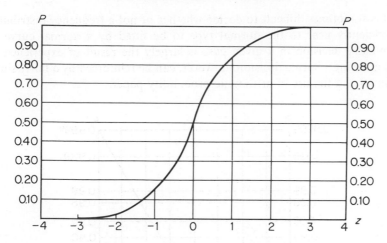

Fig. 7–28. A cumulative normal curve.

These become the ordinates of these z's. The resulting curve has the sigmoidal form shown in Figure 7–28.

7.15. Probability Paper

Imagine that the curve of Figure 7–28 were plotted on a sheet of rubber fastened to a wall by means of a rigid rod extending along the horizontal line, $P = 0.50$. Next let us stretch this rubber, upward above the line, downward below the line, in a nonuniform manner so that the curve becomes a straight line passing obliquely through the point whose coordinates are $z = 0$, $P = 0.50$. (See Figure 7–29.) Obviously slight stretchings only are necessary in the neighborhood of the horizontal line $P = 0.50$ and increased stretching is required as we move farther and farther above and below this line. The net effect is to transform the uniform scale of probabilities into a nonuniform scale as shown in Figure 7–29. The nearer the probabilities are to 0 or to 1, the greater the scale interval.

It is clearly possible to fit a theoretical cumulative normal curve to a sample distribution that appears to be drawn from a normal population by assuming $\bar{x} = \mu$ and $s = \sigma$. By the transformation just described, the cumulative curve is transformed to a straight line. When such an alteration of the axis has taken place, the resulting coordinate system with uniform (arithmetic) scale along the horizontal or z-axis and nonuniform scale along the vertical or P-axis, is used to construct *arithmetic probability paper*. We note here that P_i equals the probability that a random value, z, chosen from a normal population is less than some given value, z_i. This may be written

$$P_i = P(z < z_i) = \int_{-\infty}^{z_i} \phi(z)\, dz.$$

It is sometimes difficult to decide whether or not a frequency distribution is sufficiently near to the normal type to be fitted by a normal curve. A preliminary decision in a given case is largely the result of experience—of good guessing. Such a decision, however, can be reinforced by a fairly simple test involving the use of arithmetic probability paper.

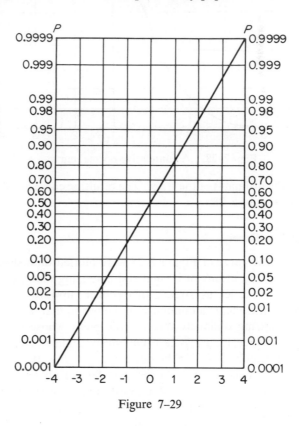

Figure 7–29

Since the area under the normal curve is unity, the partial areas, P_i, represent the *percentage cumulative frequencies* of a normal curve. For example, if we refer to Figure 7–18, we find that about 2 per cent of the normally distributed z's have values less than -2, about 16 per cent have values less than -1, 50 per cent less than 0, and so on.

We illustrate the use of the paper with the aid of Table 7–2 and Figure 7–30. Table 7–2 contains the familiar data of head lengths. Inasmuch as cumulative frequencies are of prime importance here, we are interested only in boundary values and not mid-values. The last column of values is found from the formula $100 \times \text{cum} f/N$. For example, the fifth number in the last column, 25.3, equals $100 \times 117/462$.

Table 7–2

PERCENTAGE CUMULATIVE FREQUENCIES
FOR HEAD LENGTHS

Boundary	Frequency f	Cum f	% Cum f
171.5		0	0
	3		
175.5		3	0.65
	9		
179.5		12	2.60
	29		
183.5		41	8.88
	76		
187.5		117	25.3
	104		
191.5		221	47.8
	110		
195.5		331	71.7
	88		
199.5		419	90.7
	30		
203.5		449	97.2
	6		
207.5		455	98.5
	4		
211.5		459	99.3
	2		
215.5		461	99.8
	1		
219.5		462	100.0

On the probability paper (Figure 7–30), we lay off the boundary points for the classes on the uniform horizontal scale. Inasmuch as the percentage cumulative frequencies 0 and 100 are of trifling importance, we omit them on the graph. The ordinates or percentage cumulative frequencies are plotted with the aid of the nonuniform vertical scale at the right. The points thus located should be clearly marked as heavy dots. If the given frequency distribution is approximately normal, the plotted points lie very nearly on a straight line. In practice, points near the extremities of the distribution are not considered very significant. In Figure 7–30, the points very nearly lie on a straight line, but the last three points, lying somewhat below the line, indicate that the right tail of the given distribution lies above that of a strictly normal distribution. Such being the case, the straight line may be constructed with the aid of a transparent ruler or taut string. When the resulting points do not reasonably approach a linear configuration, the distribution is not considered to be normal.

In the case of a fairly normal-appearing distribution, the approximating straight line can be used to make graphic estimates of Q_1, Q_3, μ, $\mu - \sigma$ and $\mu + \sigma$, (hence σ) for these are represented by points marking off partial areas of 0.25, 0.75, 0.50, 0.16, and 0.84 respectively, on the x-axis of the normal curve. Hence, if we locate on the horizontal scale of the probability paper the marks corresponding to these per cents (right-hand vertical scale) on the straight line, we are able to estimate these parameters. One may also estimate the graduated frequencies of a distribution. (Section 7.12.)

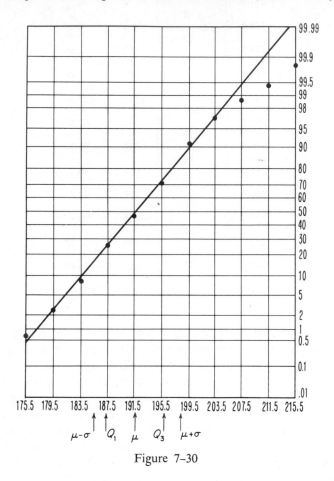

Figure 7–30

Of course, the values of statistical measures obtained by the methods just described depend upon the position of the particular line selected as the one best fitting the plotted points. Inasmuch as this is purely a matter of visual estimation, too much reliability should not be placed on the numerical results obtained. Nevertheless, probability paper furnishes us a simple, quick method of testing for normality and, in approximately normal cases, of getting rough estimates of the chief statistical constants. In dealing with discrete data, one must use artificial boundary values.

EXERCISES

By using Table C of the Supplement, find $\phi(z)$ for the values of z given in Exercises 1–3.

1. (a) 0.50; (b) 0.25; (c) -1.18; (d) 2.57.

2. (a) 1.78; (b) 0.32; (c) -2.70; (d) -1.35.

3. (a) 3.12; (b) -1.04; (c) 0.66; (d) -2.86.

By using Table D of the Supplement, find the areas (probabilities) indicated in Exercises 4–7.

4. a) $\displaystyle\int_{0.40}^{\infty} \phi(z)\, dz;$ b) $\displaystyle\int_{-\infty}^{-1.38} \phi(z)\, dz;$ c) $\displaystyle\int_{0.25}^{0.92} \phi(z)\, dz.$

5. a) $\displaystyle\int_{2.71}^{\infty} \phi(z)\, dz;$ b) $\displaystyle\int_{-\infty}^{-2.75} \phi(z)\, dz;$ c) $\displaystyle\int_{2.00}^{2.67} \phi(z)\, dz.$

6. a) $\displaystyle\int_{0.35}^{1.09} \phi(z)\, dz;$ b) $\displaystyle\int_{1.23}^{2.36} \phi(z)\, dz;$ c) $\displaystyle\int_{-0.46}^{0.46} \phi(z)\, dz.$

7. a) $\displaystyle\int_{2.10}^{3.06} \phi(z)\, dz;$ b) $\displaystyle\int_{1.00}^{\infty} \phi(z)\, dz;$ c) $\displaystyle\int_{-2.65}^{-1.65} \phi(z)\, dz.$

By means of inverse interpolation in Table D of the Supplement, find the values of z_1 corresponding to the areas given in Exercises 8–10.

8. a) $\displaystyle\int_{z_1}^{\infty} \phi(z)\, dz = 0.2819;$ b) $\displaystyle\int_{z_1}^{\infty} \phi(z)\, dz = 0.4625.$

9. a) $\displaystyle\int_{-\infty}^{-z_1} \phi(z)\, dz = 0.2438;$ b) $\displaystyle\int_{-z_1}^{z_1} \phi(z)\, dz = 0.9940.$

10. a) $\displaystyle\int_{-z_1}^{0} \phi(z)\, dz = 0.2589;$ b) $\displaystyle\int_{0}^{z_1} \phi(z)\, dz = 0.4692.$

11. With the aid of Table C (ordinates), plot carefully a normal curve on a full sheet of rectangular coordinate paper. Use values of z from -3 to 3 at intervals of 0.2.

12. With the aid of Table D (areas), plot on a full sheet of rectangular coordinate paper a cumulative normal curve by using intervals of 0.2 from $z = -3$ to $z = 3$.

In the following exercises assume all distributions to be normal.

13. If the scores in a certain test have $\mu = 500$ and $\sigma = 100$, what per cent of the scores are between 400 and 600?

14. The mean systolic blood pressure of men 20–24 years of age is 123 with a standard deviation of 13.7. What is the probability that a man of this age group, selected at random, has a blood pressure (a) above 140? (b) below 110?

15. It is desired to form a company of soldiers 6 feet or more tall. If the mean height for the regiment is 68 inches with a standard deviation of 2.5 inches, how many such soldiers may be expected to be found in this regiment of 1200 men?

16. If the mean stature of college women were 5 feet 6 inches and the standard deviation, 2.40 inches, what percentage of college women would have statures between 5 feet 2 inches and 5 feet 8 inches?

17. If the thicknesses of a certain type of washer have $\mu = 1.95$ mm and $\sigma = 0.12$ mm, how many washers in 1000 should have thicknesses between 1.80 and 2.10 mm?

18. In a certain community the mean per cent of income saved by families was 8.32 with a standard deviation of 1.94. What percentage of the families saved more than 10 per cent of their income?

19. A man drives to work every day and finds that the time per trip from his home to his office shows $\mu = 35.5$ minutes and $\sigma = 3.11$ minutes. If he leaves his home each day at 8:20 and must be in his office at 9, how many times per year should he expect to be late? Assume 240 trips per year.

20. For cats the mean lethal dose of a certain tincture of digitalis is 13.40 cc and $\sigma = 0.845$. What percentage of the cats would you estimate to die for a dose less than 12.00 cc?

21. What is the value of z corresponding to the 70 percentile? (Three decimal places.)

22. If the range (mean horizontal distance traveled) of a certain gun at a given elevation is 2560 yards with a standard deviation of 30 yards, how many shots in 100 should be "over" or "short" by more than 70 yards? If a burst within 10 yards of the target is considered a hit, how many hits should there be if the gun is "on the target"?

23. An examination has a mean score of 500 and $\sigma = 100$. The top 75 per cent of candidates taking this examination are to be passed. What is the lowest passing score?

24. The tolerance (acceptance) limits for the diameters of ball bearings produced by a machine are $\mu \pm k\sigma$. What percentage of the bearings will be accepted if (a) $k = 2.0$? (b) $k = 2.5$? (c) the lower limit corresponds to $k = 2.0$ and the upper, to $k = 2.5$?

25. The diameters of tubes made by a machine show a mean of 9.80 mm and a standard deviation of 0.536 mm. All tubes less than 9.00 mm in diameter are rejected. What percentage are rejected?

26. The blood pressures of U.S. soldiers have $\mu = 127$ and $\sigma = 14.0$. It is desired to separate from a regiment those soldiers whose blood pressures belong to the highest ten per cent. At what blood pressure should the separation be made?

27. Problems of the following type arise in anthropology (matching bones) and in industry (assembling parts). A characteristic x has $\mu = 50$ and $\sigma = 10$. A characteristic y, independent of x, has $\mu = 20$ and $\sigma = 5$. In a random sample of one item from each population, what is the probability of obtaining a value of $x > 60$ at the same time that $y < 17$?

28. Extensive records of loss of weight by evaporation of a certain packaged product show a mean loss of 6.45 grams with a standard deviation of 1.30 grams. If two packages were selected at random from a lot of them, what is the probability that both of them would show a loss of more than 8.00 grams each?

29. For the data of the preceding problem, what is the probability that, if five packages were selected, at least one package would show a loss of weight of more than 8.00 grams?

In the following exercises use the normal approximation to the binomial distribution. If binomial probability tables are available (see References 21 and 24) check your answers by them.

30. In a certain mathematics examination the number of failures over a period of years constituted 10 per cent of the candidates. In 1959 there were 60 failures in a group of 517 who took this examination. If we take $\alpha = 0.02$, should we attribute this result to chance?

31. In general 80 per cent of the patients inoculated with a certain serum recover. If only 45 out of 75 patients over 60 years of age recover after inoculation, could we accept the hypothesis that older persons react differently to this serum? Assume $\alpha = 0.02$.

32. If a pair of dice were tossed 720 times and showed the sum 6 just 80 times, would you consider the outcome unusual? At what significance level?

33. We are to test the hypothesis that thumb tacks, when dropped on the floor, land point up 60 per cent of the time. One hundred of them are dropped and 53 land point up. If $\alpha = 0.05$, and we use a two-tail test, should we accept or reject the hypothesis?

34. In New England 40 per cent of the population has been found to belong to blood group A. (a) Write, but do not compute the probability that exactly 90 persons in a New England community of 200 belong to this group. (b) Compute this probability by means of the normal approximation.

35. If the sum 5 appeared 33 times when a pair of dice is tossed 450 times, would you consider the outcome unusual at the 5 per cent level?

36. As a rule 8 per cent of the students in a certain college are elected to Phi Beta Kappa. What is the probability that at least 15 of the 120 veterans graduating are elected?

37. In an unfair game of chance, the probability that a certain man wins is $3/5$. If he makes 100 bets of one dollar each, what is the probability that his profit is at least 10 dollars?

38. How many times must you toss a penny in order that the chance of getting a deviation of more than 10 from the mean number of heads is $\frac{2}{5}$?

39. The chance of guessing a certain card correctly is $\frac{1}{5}$. In 800 trials there were 207 correct guesses. Was this an unusual result? Let $\alpha = 0.05$.

40. Compute to three decimal places the probability of obtaining less than four heads when a penny is tossed 10 times by means of (a) the binomial probability formula and (b) the normal approximation.

41. If two guinea pigs, one of pure black race and the other of pure white race are mated, the probability that an offspring of the second generation is pure white is $\frac{1}{4}$. What is the probability that among 400 such offspring more than 115 are pure white?

42. The probability of winning a game of "craps" is 0.495. If a player won 27 games out of 50, what could you say about this result?

43. Assume that when *macrosiphoniella sanborni* (the chrysanthemum aphis) are sprayed with a concentration of four milligrams per liter of rotenone, one-third of the insects die. Find the mean and standard deviation for the number dying when samples of 50 of the same insects are sprayed. What conclusion would you draw if just half of a sample of 50 died?

44. A sample of 1000 U.S. soldiers showed a mean systolic blood pressure of 127 and a standard deviation of 14.0. To the histogram for the data a normal curve is to be fitted. Compute the graduated frequency for the class interval 140–145 corresponding to the observed frequency 79.

45. The frequency distribution of a group of index numbers follows:

Index nos.	Frequency
80– 85	2
85– 90	4
90– 95	5
95–100	7
100–105	19
105–110	27
110–115	22
115–120	9
120–125	4
125–130	1
	100

Graduate, by the method of areas, the frequencies for 90–95, 95–100, and 100–105, given $\bar{x} = 107$, $s = 8.85$.

Fit a normal curve to each of the frequency distributions in Exercises 46–48. Graduate the data as directed by your instructor.

46. Thicknesses of washers. (Ex. 3.12.)

47. Brain-weights of Swedish males. (Ex. 3.6.)

48. Sizes of shoes worn by college girls. (Ex. 3.9.)

By means of arithmetic probability paper, test the distributions in Exercises 49–51 to see if they appear to be of normal type. When they are, estimate from your graph the values of μ, Q_1, Q_3, Q, and σ. Where possible, compare these values with those previously computed by other methods.

49. Thicknesses of washers. (Ex. 3.12.)

50. Brain-weights of Swedish males. (Ex. 3.6.)

51. Sizes of shoes worn by college girls. (Ex. 3.9.)

52. Two large samples were suspected to be nonnormal. The first showed $a_3 = -1.56$ and $a_4 = 2.10$. The second showed $a_3 = 0.10$ and $a_4 = 3.17$. What can you say about them?

53. The weights of adult males between 30 and 40 years of age for Race X and Race Y were compared by means of two random samples of about 1000 individuals each. The weights in pounds for Race X showed $\bar{x} = 142$, $s = 15.2$, $a_3 = 0.68$, and $a_4 = 2.85$; the corresponding statistics for Race Y were $\bar{y} = 138$, $s = 10.1$, $a_3 = 0.40$, and $a_4 = 3.19$. Compare the weight characteristics of these two races as fully as you can.

54. With the aid of probability paper, graduate the distribution of Exercise 30.

55. In Example 4 of Section 7.10, calculate β if $p = 0.42$ were true.

56. For the data of Exercise 33 let us test $H_0:p_0 = 0.60$ with $\alpha = 0.05$ and $N = 100$, against $H_1:p_1 < 0.60$. (a) What is the critical rejection number of times for the thumb tacks to land point up? (b) If $H_1:p_1 = 0.55$, what is the value of β?

INFERENCES
FROM SAMPLE MEANS

8

"Everywhere one observes the unfortunate habit of generalizing,
without demonstration, from special cases."
NIELS ABEL, 1826

8.1. Introduction

Probably the largest area of statistical inquiry deals with the drawing of conclusions concerning a population, from a sample selected from it. Such conclusions will never be certain; a probability of being correct will be attached to each inference. For this reason we often speak of this aspect of statistics as *uncertain inference*. The fact that useful inferences can be made from small samples, even as small, say, as three or four observations, is often difficult for persons to believe. As this book progresses, the reasons therefore should become clear.

8.2. The Distribution of a Sample Mean

Suppose that from the population of adult American males of age 21 years, we select, independently of one another, a very large number of random samples of a fixed size, N, say $N = 25$. For each sample we could calculate a variety of statistics on stature—the arithmetic mean, $\bar{x}$, the standard deviation, s, the median, x_m, and so on. Let us fix our attention on the means. We could construct a frequency table and a histogram of these sample means and study their distribution. We could calculate the mean of these means, the standard deviation of these means, the median of these means, and so forth. What might we discover? One can guess at some of the results to be found, but on others, the guesses would likely be incorrect. Let us see what would happen.

(1) The histogram or frequency polygon of means would appear to indicate a normal distribution of them.

(2) The mean of the means would be very close to the mean, μ_x, of the population.

(3) The standard deviation of the means, $s_{\bar{x}}$, would be very much smaller than the standard deviation, σ_x, of the population. In fact, for our distribution of means of samples of 25, $s_{\bar{x}}$, would come out to about $\frac{1}{5} \sigma_x$.

(4) The median of the means would be very close to the median of the population.

The properties listed above originate from the nature of the frequency function for $\bar{x}$. It is proved by the methods of mathematical statistics that the means, $\bar{x}$, of samples of N, each drawn from the same normal population, tend to be normally distributed, with a population mean equal to the mean of the population of x, and with a variance equal to $1/N$th the variance of the population. In symbols,

$$\mu_{\bar{x}} = \mu_x, \tag{8.1}$$

$$\sigma_{\bar{x}}^2 = \frac{\sigma_x^2}{N}. \tag{8.2}$$

If the population is normal, it can be proved mathematically that the sample mean, $\bar{x}$, has precisely a normal distribution. In this connection it is worthwhile to state a very important theorem of statistics.

THE CENTRAL LIMIT THEOREM. *The mean, $\bar{x}$, of a sample of N drawn from any population (continuous or discrete) with mean, μ, and finite variance, σ^2, will have a distribution that approaches, as N becomes infinite, the normal distribution with mean, μ, and variance, σ^2/N.*

The proof of this theorem demands advanced mathematical methods, and cannot be given here.

It will be noted that this theorem implies that for large N, $\bar{x}$ has practically a normal distribution, even if the population is not normal. Whether it is normal or not, Equations (8.1) and (8.2) hold. In the work of this book we shall assume all the populations from which the samples are drawn to be approximately normal, hence we may assume $\bar{x}$ to have a normal distribution.

The quantity obtained by taking the positive square root of (8.2)

$$\sigma_{\bar{x}} = \frac{\sigma_x}{\sqrt{N}}, \tag{8.3}$$

is often called the *standard error of the mean*. The relationship of the frequency functions of $\bar{x}$ for samples of 9 and 25 to the parent normal population is illustrated in Figure 8–1.

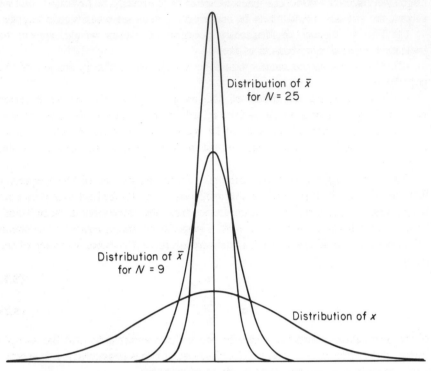

Fig. 8–1.   The normal distribution of x and the distributions of $\bar{x}$ for
samples of 9 and 25.

8.3.   Applications

In most practical problems $\bar{x}$ has essentially a normal distribution, therefore we may test a hypothesis about a population mean, μ_x, by drawing a sample of N from the population and calculating its mean, $\bar{x}$. In the type of problem now under discussion we assume that the population standard deviation, σ_x, is known. The standard variable now becomes

$$z = \frac{\bar{x} - \mu_{\bar{x}}}{\sigma_{\bar{x}}}, \quad \text{where} \quad \mu_{\bar{x}} = \mu_x, \quad \text{and} \quad \sigma_{\bar{x}} = \frac{\sigma_x}{\sqrt{N}};$$

hence

$$z = \frac{(\bar{x} - \mu_x)\sqrt{N}}{\sigma_x}. \tag{8.4}$$

EXAMPLE 1.   Suppose that the mean length of life of 40-watt electric light bulbs is guaranteed to be at least 1000 hours and that the standard deviation is 200 hours. A sample of 16 is drawn from a shipment of bulbs and shows a mean length of life of 910 hours. Does this indicate that the shipment does not meet the guarantee and should be rejected?

Let the null hypothesis, H_0, be $\mu \geqq 1000$, the alternative hypothesis, H_1 be $\mu < 1000$, and $\alpha = 0.05$. [See Section 6.12(4).] Here the buyer is primarily

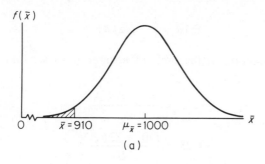

(a)

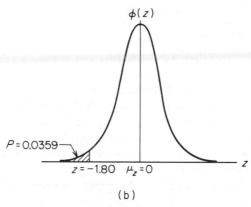

(b)

Figure 8–2

interested in *not* accepting a lot of bulbs whose lives have a mean *below* that guaranteed, 1000 hours. By (8.4)

$$z = \frac{(910 - 1000)\sqrt{16}}{200} = -1.80.$$

The probability that a mean of 16 is 910 or less (left-tail area),

$$P(\bar{x} \leqq 910) = \int_{-\infty}^{-1.80} \phi(z)\, dz = 0.0359.$$

[See Figures 8–2(a), (b).] Since $P < 0.05$, the sample indicates that the shipment contains too many bulbs with lives below 1000 hours and should be rejected.

EXAMPLE 2.   On a nation-wide examination the scores showed $\mu = 72$ and $\sigma = 10$. How large a sample of candidates from the University of X must be taken, in order that there be a 10 per cent chance that its mean score is less than 70?

Since $P(\bar{x} < 70)$ is to equal 0.10, we seek the value of z_1 such that

$$0.10 = \int_{-\infty}^{z_1} \phi(z)\, dz.$$

By inverse interpolation in the table of normal areas we find that z_1 must equal -1.28.

Since $$z = \frac{(\bar{x} - \mu)\sqrt{N}}{\sigma_x},$$

$$-1.28 = \frac{(70 - 72)\sqrt{N}}{10},$$

whence $$N = 40.96.$$

Therefore we must take a sample of 41.

EXAMPLE 3.   Suppose that we are testing the breaking strength, x, of a certain type of cord by means of a random sample of 9 pieces of cord. Let $\sigma = 5$ pounds and set up the hypothesis, H_0, that the population mean breaking strength, $\mu_0 \geq 50$ pounds. Assume that the alternative, H_1, to this hypothesis is $\mu < 50$. Choose the type I error to be (say) 0.05.

(a) Find the rejection and acceptance regions for H_0.

(b) If the alternative hypothesis, H_1, were true with $\mu = 45$, what would be the size of the Type II error, β.

It would be wise for the reader, at this point, to review Section 6.11.

(a) By the use of Formula (8.4) we find that, corresponding to a left-tail area of 0.05, $z = -1.645$. Then

$$-1.645 = \frac{(\bar{x} - 50)\sqrt{9}}{5},$$

whence $\bar{x} = 47.3$. Thus there is a probability of not more than 0.05 that a sample of 9 will yield by chance, an $\bar{x}$ as low as 47.3 or lower when $\mu \geq 50$. (Figure 8–3). If then our sample yields an $\bar{x}$ lower than 47.3 we reject H_0 and accept as an alternative the conclusion that $\mu < 50$. The interval $\bar{x} < 47.3$ constitutes the *critical* or *rejection region* for the hypothesis $H_0 : \mu \geq 50$. The interval, $\bar{x} > 47.3$ is the *acceptance region* for H_0. If, however, H_0 were true, repeated tests of samples of 9 would produce $\bar{x}$'s lying in the rejection region 5 per cent of the time and we would then be committing a Type I error of size $\alpha = 0.05$. Also, if H_0 were true, the $\bar{x}$'s would lie in the acceptance region 95 per cent of the time and this is the percentage of times that we would be making a correct decision.

(b) If H_0 were false then the true mean would be less than 50. The magnitude of the Type II error, β, will depend upon the true value of μ which we shall now assume to be $\mu_1 = 45$. How often would we accept

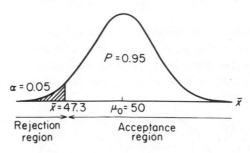

$P = 0.95$

$\alpha = 0.05$

$\bar{x} = 47.3$ $\mu_0 = 50$ $\bar{x}$

Rejection region   Acceptance region

Figure 8–3

$H_0: \mu \geq 50$ when $H_1: \mu_1 = 45$ is true? The chance of obtaining an $\bar{x} > 47.3$ when $\mu = 45$ is found thus:

$$z = \frac{(47.3 - 45)\sqrt{9}}{5} = 1.38,$$

and the corresponding probability is 0.0838. (See the horizontally shaded area in Figure 8–4.) Thus β is about 0.08; in other words, we shall accept the hypothesis that $\mu \geq 50$ when actually $\mu = 45$, about 8 per cent of the

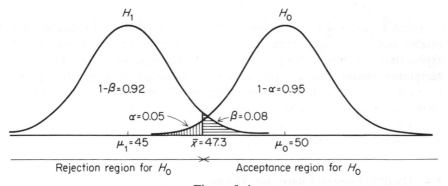

H_1 H_0

$1 - \beta = 0.92$ $1 - \alpha = 0.95$

$\alpha = 0.05$ $\beta = 0.08$

$\mu_1 = 45$ $\bar{x} = 47.3$ $\mu_0 = 50$

Rejection region for H_0 Acceptance region for H_0

Figure 8–4

time. Note also that we shall accept the alternative $\mu < 50$, that is, reject H_0, about 92 per cent of the time when $\mu_1 = 45$, and this would be the correct decision.

As an additional illustration suppose that $\mu_1 = 48$ instead of 45. How often would we accept $H_0: \mu_0 \geq 50$?

The chance of obtaining an $\bar{x} > 47.3$ when $\mu = 48$ is found in the same manner as before.

$$z = \frac{(47.3 - 48)\sqrt{9}}{5} = -0.42,$$

whence the probability, β, sought is 0.6628 (Figure 8-5.) Thus we shall accept the hypothesis that $\mu_0 \geq 50$ when it is false about 66 per cent of the time provided the true mean is 48. Also we shall accept $H_1 : \mu < 50$ or, what is equivalent, reject $H_0 : \mu \geq 50$ about 34 per cent of the time when $\mu = 48$.

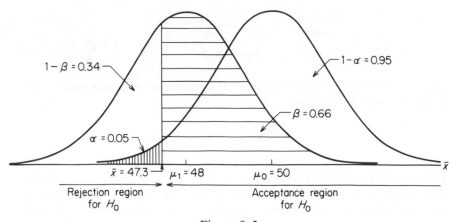

Figure 8-5

From Figure 8-4 it will be observed that the farther μ_1 is below μ_0, the smaller the chance of committing a Type II error. Intuitively we should expect that the smaller the μ_1's are, the less likely we are to get $\bar{x}$'s in the acceptance region, that is, large enough to cause us to accept H_0. From Figure 8-5 μ_1 is closer to μ_0 so that it is more difficult to discriminate between them by means of a statistical test. The error, β, becomes larger as μ_1 approaches μ_0. Considerations such as these lead us to the concept of the *power function*. (See Section 8.12.)

8.4. Quality Control Charts for Means

In many industrial processes it is desirable to test at frequent intervals the quality of the products manufactured. If a specified measurement of a given article is to fall within certain desired limits of precision, random samples of items are taken from time to time, measured, and the means of the samples calculated. These values are plotted on a control chart (Figure 8-6) and studied for unusual behavior.

Assume that previous experience has established the mean, μ, and the standard deviation, σ_x, of the population of measurements. Then, for samples of size N, $\sigma_{\bar{x}} = \sigma_x/\sqrt{N}$. A horizontal line is drawn to represent μ and parallel lines are drawn at distances of $3\sigma_{\bar{x}}$ above and below it. As the sample means are reported, they are plotted as dots on the chart. Since the chance

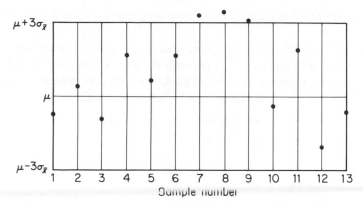

Figure 8–6

that an $\bar{x}$ lies outside the interval $\mu - 3\sigma_{\bar{x}}$ to $\mu + 3\sigma_{\bar{x}}$ is about 0.0026, any dot falling outside of the three-sigma limits is a strong indication that something is wrong in the manufacturing process, and the cause is immediately sought. Later sampling will indicate whether or not the process has been brought back under control.

In Figure 8–6, samples 7, 8, and 9 indicate lack of control, and sample 10 indicates that control of quality was re-established. The fact that most of the dots were above the central line might, in certain industrial experience, indicate a trend of importance.

8.5. A General Theorem

An important fundamental problem arising in mathematical statistics is this: Given two or more statistics (means, proportions, standard deviations, etc.) obtained from samples drawn from the same or from different populations whose frequency functions are known, what is the frequency function of some combination of these statistics?

One general result of immediate importance is the following:

If x and y are two variables with independent frequency functions having means μ_x and μ_y and variances σ_x^2 and σ_y^2 respectively, then the population mean of their difference, x — y, (or sum, x + y) is the respective difference (or sum) *of their population means, and the population variance of their difference (or sum) is the sum of their population variances, that is,*

$$\mu_{x-y} = \mu_x - \mu_y, \tag{8.5}$$

$$\mu_{x+y} = \mu_x + \mu_y, \tag{8.6}$$

$$\sigma_{x-y}^2 = \sigma_{x+y}^2 = \sigma_x^2 + \sigma_y^2. \tag{8.7}$$

It is interesting that the last formula holds for either $x - y$ or $x + y$. Furthermore,

If x and y are normally distributed then $x - y$ and $x + y$ are also.

If $\bar{x}$ and $\bar{y}$ are means of sample sizes N_x and N_y respectively, which are normally and independently distributed, then by the preceding theorems their difference, $\bar{x} - \bar{y}$, is also normally distributed with

$$\mu_{\bar{x}-\bar{y}} = \mu_{\bar{x}} - \mu_{\bar{y}} = \mu_x - \mu_y, \tag{8.8}$$

$$\sigma_{\bar{x}-\bar{y}}^2 = \sigma_{\bar{x}}^2 + \sigma_{\bar{y}}^2$$

$$= \frac{\sigma_x^2}{N_x} + \frac{\sigma_y^2}{N_y}, \tag{8.9}$$

by (8.2). It follows that a hypothesis on $\mu_x - \mu_y$ can be tested by means of the standard variable

$$z = \frac{(\bar{x} - \bar{y}) - \mu_{\bar{x}-\bar{y}}}{\sigma_{\bar{x}-\bar{y}}}$$

$$z = \frac{(\bar{x} - \bar{y}) - (\mu_x - \mu_y)}{\sqrt{\dfrac{\sigma_x^2}{N_x} + \dfrac{\sigma_y^2}{N_y}}}. \tag{8.10}$$

Example. Suppose that 64 senior girls from College A and 81 senior girls from College B had mean statures of 68.2 inches and 67.3 inches respectively. If the standard deviation for statures of all senior girls is 2.43 inches, is the difference between the two groups significant?

Since mean statures are normally distributed, the difference $\bar{x} - \bar{y}$ is also, and Formula (8.10) is applicable. We take as our null hypothesis that there is no difference in mean height between the senior girls of Colleges A and B; in symbols

$$H_0 : \mu_x - \mu_y = 0.$$

Let us take as our alternative hypothesis,

$$H_1 : \mu_x - \mu_y \neq 0$$

so that a two-tail test is called for. We also assume that

$$\sigma_x = \sigma_y = 2.43$$

and take $\alpha = 0.05$.

Then from (8.10)

$$z = \frac{(68.2 - 67.3) - 0}{\sqrt{\frac{(2.43)^2}{64} + \frac{(2.43)^2}{81}}} = 2.21.$$

Since $2\int_{1.96}^{\infty} \phi(z)\, dz = 0.05$, the rejection region for H_0 consists of the two intervals $z < -1.96$ and $z > +1.96$. The number 2.21 corresponds to a point in the right-tail rejection interval so the hypothesis, H_0, is refuted; the girls from the two colleges differ significantly in their statures.

8.6. Remarks

It is important to note that the method of the preceding section as well as that of Section 8.3 requires that the population variances be known. In many cases they are not known and must be estimated from the sample. If the sample is large, say $N \geq 50$, the sample variance s^2 may be used in place of the unknown parameter, σ^2, without too much risk. When N is small, it is necessary to employ the t-distribution which will be explained in the next section.

Example. Two sections of 54 and 67 students each took the same examination in statistics. The first section had a mean score of 73.1 with a standard deviation of 11.2; the second had a mean score of 76.6 with a standard deviation of 13.0. Can the difference in mean scores be attributed to chance? Let $H_0: \mu_x - \mu_y = 0$ and $H_1: \mu_x - \mu_y \neq 0$. Let $\alpha = 0.05$.

Since both sample numbers are moderately large, assume $\sigma_x = s_x = 11.2$ and $\sigma_y = s_y = 13.0$. Then by (8.10)

$$z = \frac{(73.1 - 76.6) - 0}{\left[\frac{(11.2)^2}{54} + \frac{(13.0)^2}{67}\right]^{1/2}} = -1.59.$$

Since the standard deviate is less numerically than 1.96, we accept H_0; there is no evidence that the difference is not due to chance.

8.7. The Student-Fisher t-distribution

The problem of testing the significance of the deviation of a sample mean from a given population mean when N is small and only the sample variance is known was first solved (1908) by W. S. Gossett, writing under the pen name of *Student*. His method was later modified by R. A. Fisher. When the population of x is normal they proved that the variable

$$t = \frac{(\bar{x} - \mu)(N - 1)^{1/2}}{s} \tag{8.11}$$

has a symmetrical, bell-shaped, but nonnormal distribution with $\mu_t = 0$. Its frequency function is of the form

$$y = C\left(1 + \frac{t^2}{n}\right)^{-\frac{n+1}{2}} \tag{8.12}$$

where the parameter $n = N - 1$ and is called the *number of degrees of freedom* (to be explained later). C is a constant depending upon n. The curve for $n = 20$, together with a normal curve, is shown in Figure 8–7. As n increases, the t-curve approaches a normal curve as a limiting form.

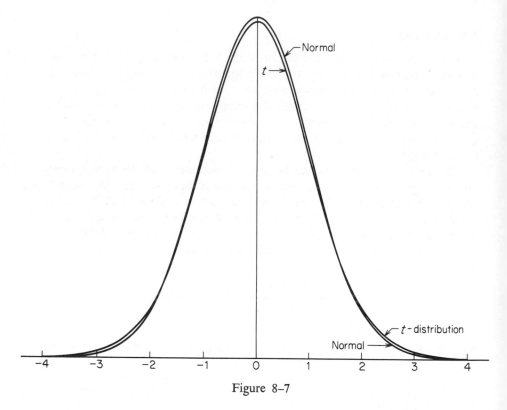

Figure 8–7

Fisher tabulated the values of t corresponding to various levels of significance for different useful values of n. A modified form of his table appears in Table E of the Supplementary Tables. From this table, for example, we find for $n = 10$ and $P = 0.01$ that $t = 2.76$. Thus (Figure 8–8) we find that

$$P(t > 2.76) = P(t < -2.76) = 0.01$$

or that $$P(|t| > 2.76) = 0.02.$$

In order to use the Student-Fisher distribution we take into account the *number of degrees of freedom* of the N variates, $x_1, x_2, \ldots, x_N$, constituting the sample from which $\bar{x}$ and s are computed. It will be desirable to explain briefly this phrase in italics.

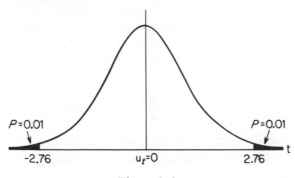

$P = 0.01$ $P = 0.01$

-2.76 $u_t = 0$ 2.76

Figure 8–8

8.8. Degrees of Freedom

Suppose that the mean score of five children on a Binet Intelligence Test is 106. What can we say about the individual scores? The answer here is obvious—very little. In fact, we can write down with the utmost freedom any four scores that occur to us, say 70, 123, 98, and 104, and then determine the fifth score, x_5 from the fact that

$$\tfrac{1}{5}(70 + 123 + 98 + 104 + x_5) = 106$$

whence, $\qquad\qquad\qquad\qquad\qquad x_5 = 135.$

It is apparent that any four scores whatsoever may be written down, but the fifth is determined by the restriction embodied in the formula for the mean,

$$\bar{x} = \tfrac{1}{5} \sum_{i=1}^{5} x_i.$$

We say, then, that the five variates have four degrees of freedom or one restriction. In the mathematical derivation of the frequency distribution of a statistic we find that the number of free variates plays an important role.

In the case of the t-distribution we may ask: Given a sample of N which determines a mean $\bar{x}$, or its deviation from the population mean, $\bar{x} - \mu$, how can t in (8.11) vary? Clearly, the variation of t, for any $\bar{x}$, depends upon that of the denominator, s. For a given $\bar{x}$ the N variates comprising the sample, $x_1, x_2, \ldots, x_N$, are free to vary and thus to determine various values of s, subject to the restriction

$$\bar{x} = \frac{1}{N} \sum_{i=1}^{N} x_i.$$

Thus, the t-distribution has $N - 1$ degrees of freedom.

8.9. Applications

EXAMPLE 1.   A machine is to turn out engine parts with axle diameters of 0.700 in. The hypothesis $H_0 : \mu = 0.700$ is to be tested against the alternative $H_1 : \mu > 0.700$, with $\alpha = 0.025$. A random sample of 10 parts shows $\bar{x} = 0.712$ inches and $s = 0.040$ inches. What does this indicate?

Solution. Since N is small and σ is not known, we use the t-test [Formula (8.11)]. For $n = 9$ and $P = 0.025$ it follows from Table E that $t = 2.262$. Then by (8.11)

$$t = \frac{(0.712 - 0.700)\sqrt{9}}{0.040} = 0.900.$$

Since $P(t > 2.262) = 0.025$, this value, 0.900, is not large enough to cause us to reject H_0. It appears that the machine is working properly.

EXAMPLE 2.   The grades of students in a certain course averaged 77 over a period of years. A class of 40 has a mean grade of 70 with a standard deviation of 9. Can this lower mean be attributed to ordinary sampling variation?

Solution. By (8.11)

$$t = \frac{(70 - 77)(39)^{\frac{1}{2}}}{9} = -4.86.$$

Inasmuch as the sample is moderately large, we may, instead of referring to the t-table, use the normal distribution. Obviously, a difference of almost five sigmas can hardly be attributed to the ordinary fluctuations of sampling.

8.10. The Difference Between Two Sample Means When the Population Variances are Unknown

When the population variances are not known but are assumed to be equal, the following methods may be used for testing the significance of the difference between two sample means obtained from two populations.

Mathematical statistics has proved that the variable

$$t = \frac{(\bar{x} - \bar{y}) - (\mu_x - \mu_y)}{\left[\left(\dfrac{N_x s_x^2 + N_y s_y^2}{N_x + N_y - 2}\right)\left(\dfrac{1}{N_x} + \dfrac{1}{N_y}\right)\right]^{\frac{1}{2}}} \tag{8.13}$$

has a t-distribution with $n = N_x + N_y - 2$ degrees of freedom. As in Section 8.5, the null hypothesis is that $\mu_x - \mu_y = 0$. If the samples are large, we revert to the method of Section 8.6.

The reason for the complex denominator in (8.13) will appear in Section 10.4 [Formula (10.7)]. Ordinarily, if the two sample variances s_x^2 and s_y^2 are not too far apart, we assume $\sigma_x^2 = \sigma_y^2$. If we are in doubt concerning this, the F-test to be studied in Section 11.12 may be employed as a test of the hypothesis, $\sigma_x^2 = \sigma_y^2$. If the two samples are drawn from populations with different variances, that is, if $\sigma_x^2 \neq \sigma_y^2$, the procedures used above may be questioned. Alternative tests are available for the case where the variances differ significantly.

Example. The mean life of a sample of 10 electric light bulbs was found to be 1456 hours with $s = 423$ hours. A second sample of 17 bulbs chosen from a different batch showed a mean life of 1280 hours with $s = 398$ hours. Is there a significant difference between the means of the two batches? Let $\alpha = 0.05$.

Let H_0 be $\mu_x - \mu_y = 0$ and let H_1 be $\mu_x - \mu_y \neq 0$. By (8.13)

$$t = \frac{(1456 - 1280) - 0}{\left[\left(\frac{(10 \times 423^2) + (17 \times 398^2)}{10 + 17 - 2}\right)\left(\frac{1}{10} + \frac{1}{17}\right)\right]^{\frac{1}{2}}} = 1.04.$$

For $t = 1.04$ and $n = 25$ we find from Table E that $P = 0.15$ approximately, so that the two-tail probability, $2P = 0.30$. We can say that there is no evidence that the two batches are significantly different in length of life. In other words, the null hypothesis, $\mu_x = \mu_y$, is not denied.

8.11. Paired Observations

It is sometimes desirable to analyze a sample of N pairs of observations, x_i and y_i, $(i = 1, 2, \ldots, N)$ for the purpose of detecting a possible difference between the two populations from which x and y have been selected. The individuals of a pair are selected because of certain common factors whose effects we wish to eliminate in the statistical experiment. The difference between the observations of each pair, $d = x - y$, becomes the variable whose statistics we calculate. Then (8.11) becomes

$$t = \frac{(\bar{d} - \mu_d)(N - 1)^{\frac{1}{2}}}{s_d}. \tag{8.14}$$

By the theorem of Section 8.5 we know that if x and y are normally distributed, then their difference $d = x - y$, is also normally distributed. The method of testing will be illustrated with the following example.

Example. Two kinds of photographic films were tested for sharpness of definition in the same camera under varying conditions. Each pair of readings shown in Columns 2 and 3 of Table 8–1 was produced under the same conditions except for difference of film. What conclusions can we draw?

Table 8–1

Pair	Readings Film X x	Readings Film Y y	Difference $d = x - y$	d^2
1	27	25	2	4
2	30	28	2	4
3	30	30	0	0
4	32	30	2	4
5	24	27	−3	9
6	26	28	−2	4
7	40	37	3	9
8	35	28	7	49
Sums	244	233	11	83

The null hypothesis is that the mean of the population of differences, $\mu_d = 0$. Let H_1 be $\mu_d \neq 0$ and choose $\alpha = 0.05$.

The calculation is conveniently arranged in Table 8–1. The mean, $\bar{d} = {}^{11}\!/_8 = 1.38$. By Formula (2.7)

$$s_d^2 = {}^{83}\!/_8 - (1.38)^2 = 8.49,$$

whence $s_d = 2.91$. Substituting these values in (8.14) we have

$$t = \frac{(1.38 - 0)\sqrt{7}}{2.91} = 1.25.$$

Entering the t-table with 7 degrees of freedom we find that $t = 1.25$ corresponds to a one-tail probability $P > 0.10$, so that for a two-tail test, $2P > 0.20$. Thus the hypothesis H_0, is accepted; there is no unusual difference between the sharpness of definition of the two films.

8.12. The Power of a Test

If we examine Figures 8–4 and 5 and note the behavior of β as μ_1 recedes to the left away from μ_0, we see that β becomes smaller and will eventually approach 0. Similarly as μ_1 moves to the right, β increases. The graph of $1 - \beta$ plotted against μ_1 for $\mu_1 < \mu_0$ is shown in Figure 8–9 and exhibits the change in the probability of accepting H_1, that is, rejecting H_0 as μ_1 varies. The probability, $1 - \beta = 1 - 0.08 = 0.92$ obtained earlier (Section 8.3) is called the *power of the test* of the hypothesis that $\mu_0 \geq 50$ with respect to the alternative, $\mu_1 = 45$; it is the probability of rejecting H_0 when the alternative H_1 is true. Similarly, the probability $1 - 0.66 = 0.34$ is the power of the test of $H_0 : \mu_0 \geq 50$ with respect to the alternative $H_1 : \mu_1 = 48$. For the case that $\mu_1 = \mu_0$, $\beta = 1 - \alpha$, whence $1 - \beta = \alpha$. The probability $P = 1 - \beta$ obviously varies with μ_1, that is, it is a function of μ_1. This function, $P(\mu)$ is called the *power function*.

If we test $H_0:\mu_0 = 50$ against $H_1:\mu_1 \neq 50$, which implies a two-tail test, we can verify that the power curve has the form shown in Figure 8–10.

Naturally we should like to keep the sizes of the two errors α and β as small as possible but we cannot reduce both simultaneously since a decrease in α causes an increase in β and vice versa. Customarily we select an

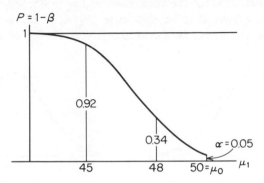

Fig. 8–9. Probability of rejecting $H_0:\mu_0 \geq 50$ in favor of $H_1:\mu_1 < 50$.

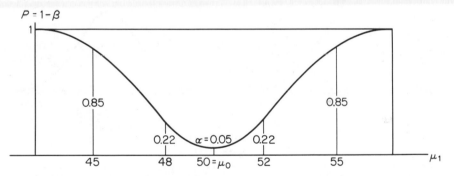

Fig. 8–10. Probability of rejecting $H_0:\mu_0 = 50$ in favor of $H_1:\mu_1 \neq 50$.

appropriate α, say, $\alpha = 0.05$ and then select a rejection region that minimizes β. In this connection it should be pointed out that an infinite number of rejection regions are, in general, possible for a given H_0 and a given α, but that only a few have practical value. Thus, if we test $H_0:\mu_0 = 50$ and choose $\alpha = 0.05$, we might select among the many rejection regions possible the following: [Figures 8–11(a)–(d)].

(a) $\bar{x} > 52.7$. Thus an excessively large $\bar{x}$ causes rejection of H_0 [Figure 8–11(a)].

(b) $\bar{x} < 47.3$. Thus an excessively small $\bar{x}$ causes rejection of H_0 [Figure 8–11(b)].

(c) $\bar{x} < 46.7$, $\bar{x} > 53.3$. Thus either an excessively small or excessively large $\bar{x}$ causes rejection of H_0 [Figure 8–11(c)].

(d) $51.9 < \bar{x} < 52.4$. This region is mathematically possible but unrealistic [Figure 8–11(d)].

If the true value of $\mu = 45$, we can calculate the power of the test for each of the four regions selected. The respective values of $P(\mu)$ follow:

(a) 0.00; (b) 0.92; (c) 0.85; (d) 0.00.

It follows then, that for the alternative, $H_1 : \mu_1 = 45$, test region (b) is the most powerful. Note that in this case $H_0 : \mu_0 = 50$, is rejected more frequently (as it should be) for a critical region to the *left* of μ_0, provided the

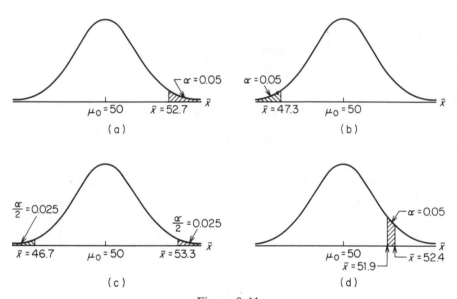

Figure 8–11

true mean, $\mu_1 < 50$. This conclusion confirms the simple intuition that $\bar{x}$'s will occur more frequently to the left of μ_0 if the true mean μ_1 lies to the left.

On the other hand, if the true mean $\mu = 55$, the power of the test for each of the same four regions becomes

(a) 0.92; (b) 0.00; (c) 0.85; (d) 0.03.

Thus test (a) is the most powerful. Again, intuitively we should expect $\bar{x}$'s to occur more frequently to the *right* of μ_0 if the true mean μ_1 lies to the right. Test (c) is not as powerful as (a) since the right rejection region lies somewhat farther down the right tail. See Exercise 43.

In industrial statistics the Type I error is often called the *producers' risk*, because the rejection of the hypothesis H_0 means that manufactured articles will not be sold. The Type II error is called the *consumers' risk*, since the consumer would then be accepting articles not meeting certain specifications.

If, instead of plotting $1 - \beta$ against μ, we plot β, we obtain an *operating-characteristic* (OC) *curve*. Figure 8–12 shows the two OC curves corresponding to the power curves of Figures 8–9 and 10. The ordinate of an OC curve, for a given μ_1, represents the probability of accepting H_0 when it is false.

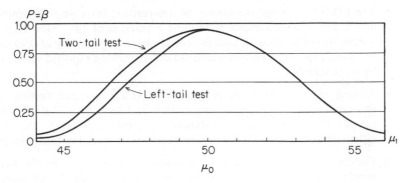

Fig. 8–12.   Operating-characteristic curves corresponding to Figures 8–9, 10.

8.13. Nonparametric Tests

It should be observed that the t-tests described usually require the assumption of normality of the underlying distribution. If this requirement is not satisfied, is in doubt, or if the data are not in the form of measurements, one may use, for example, the rank-sum test of Section 15.7 or, for paired observations, the sign test of Section 15.4.

EXERCISES

In Exercises 2–21 assume normality for the distributions.

1. If the scores of students in Graduate Record Examinations show a standard deviation of 100, what will the standard deviation be for the mean scores of samples of (a) 16? (b) 25? (c) 100?

2. A certain test has been administered to thousands of pupils with a mean score of 80 and a standard deviation of 6. The same test given to 64 pupils of race R showed a mean of 78. Are scores for this race substantially lower than those in general? Let $\alpha = 0.05$.

3. Experience shows that a fixed dose of a certain drug causes an average increase of pulse rate of 10 beats per minute with a standard deviation of 4. A group of 9 patients given the same dose showed the following increases:

$$13, \quad 15, \quad 14, \quad 10, \quad 8, \quad 12, \quad 16, \quad 9, \quad 20.$$

Is there evidence that this group is different in response to the drug? ($\alpha = 0.05$.)

4. A certain drug is found, on the average, to lower the systolic blood pressure by 12.3 with $\sigma = 4.70$. When administered to a group of 16 selected patients, it was found that their blood pressures were changed by the following amounts: $-15, -7, -3, -12, -20, -9, -5, -9, -11, -10,$ $0, -7, -18, +5, -21, -15$. Test the hypothesis that this type of patient reacts no differently with respect to the drug than other patients in general.

5. The I.Q.'s for a large population of five-year-old boys were determined and found to have a mean of 110 and a variance of 144. How large a sample would we have to take in order to have a probability of 0.90 that the mean I.Q. of the sample would not differ from the expected value 110, by more than five?

6. Assume that the mean specific gravity of U.S.P. coal tar available to pharmacists for use in coal tar preparations is 1.185 with $\sigma = 0.012$. Ten samples of nonofficial coal tar showed a mean S.G. of 1.172. Do these differ significantly ($\alpha = 0.05$) with respect to S.G. from the U.S.P. mean?

7. For the ampuls of Arsenical I manufactured by a company, the mean weight is 145.3 mg with $\sigma = 4.82$ mg. How large a sample of such ampuls should we take in order that we may be 95 per cent sure that the mean weight does not exceed 147.0 mg?

8. A manufacturer requires cotton thread with a mean breaking strength of 6.50 ounces and a $\sigma = 0.25$. He tests a new lot of thread by means of a sample of 16 pieces and finds $\bar{x} = 6.65$ ounces. Does this indicate that the lot is stronger than required?

9. Assume a population to have $\sigma = 10$. Choose $\alpha = 0.05$. Find the rejection and acceptance regions for the hypothesis $\mu \leq 50$. What is the Type II error, β, when the true mean is 53? 55? Assume $N = 25$.

10. Assume that the mean weight of some newly compounded tablets is to be equal to 4.00 grams and that the standard deviation is expected to be 0.20 grams. Samples of 16 tablets are to be tested and the rejection region for $H_0: \mu_0 \leq 4.00$ is $\bar{x} > 4.10$. (a) Find the value of α. (b) Find the value of β if $H_1: \mu_1 = 4.10$.

11. We are to test the hypothesis $H_0: \mu_0 \leq 150$, by means of a sample of 25. Assume $\sigma = 20$ and $\alpha = 0.07$. (a) Find β when $H_1: \mu_1 = 160$. (b) Find the power of this test.

12. Assume that pulse rates for American males of age 25 have $\sigma = 9.00$. For samples of 100, with $\alpha = 0.03$ (a) find the acceptance region for the hypothesis that $\mu_0 = 72.0$ when the alternative $\mu_1 \neq 72.0$. (b) Find β if the alternative hypothesis is $\mu_1 = 71.0$. (c) Find the power of the test for the alternative in (b).

13. In Example 3 of Section 8.3 replace the hypothesis $\mu_0 \geq 50$ which calls for a one-tail test by the hypothesis $\mu_0 = 50$ which demands a two-tail test for the alternative $H_1: \mu_1 \neq 50$. Let $\alpha = 0.05$ and show that the rejection

region consists of the two intervals $\bar{x} < 46.7$ and $\bar{x} > 53.3$. If $\mu_1 = 45$ show that $\beta = 0.15$. This choice of critical region produces a test not as good as the earlier one for $\mu_0 \geq 50$, because the same α corresponds to a larger β.

14. A sample of 49 tablets is to be tested for weight. Assuming $\sigma = 0.25$ grams and $\alpha = 0.10$, find the acceptance region for the hypothesis, $\mu = 2.50$ grams when the alternative is that $\mu > 2.50$ grams. Find the power of the test when the true mean $\mu = 2.60$ grams. What does your answer mean?

15. In the production of telescoping whip antennas for military portable radios, the length of the collapsed antenna proved to be a critical dimension. If the mean length is to be 14.980 inches and the standard deviation is estimated to be 0.0150 inches, construct a control chart for the means of successive samples of 5 each, which follow, and discuss its evidence.

14.982, 14.994, 14.979, 14.984, 14.985, 14.995, 14.993, 14.980,
14.978, 14.998, 15.011, 14.993, 14.985, 14.998, 15.007, 14.986.

16. Control charts may be used to measure personnel performance. Construct a control chart with two-sigma limits for the weekly means, given below, of the number of parts inspected per day for successive weeks by inspector A, given that the mean number of parts inspected per inspector per day is 811 and the standard deviation is 286. Assume a working week of 5 full days, with the first week ending February 5, 1960. Suggest reasons for the variations that appear on your chart and discuss their effect on the manufacturing costs.

798, 964, 1002, 790, 901, 957, 1092, 911, 836,
1310, 1144, 965, 788, 783, 482, 400, 480, 571.

17. The heights of the 16-year-old boys in two schools of the same city have been sampled. Fifty boys from School X have a mean height of 66.00 inches and an equal number from School Y have a mean height of 65.50 inches. If σ for the heights of all 16-year-old boys is 2.00 inches can we say that the difference in heights is significant at the 5 per cent level?

18. The systolic blood pressures of a group of 60 patients showed $\bar{x} = 140$, $s_x = 10$, and x_m (median) $= 141$. A second group of 60 showed $\bar{y} = 145$, $s_y = 13$, and $y_m = 138$. Compare the two groups and give reasons for your statements.

19. The total nitrogen (N) content (mg per cc) of rat blood plasma was determined for a group of 60 rats of age 50 days and for a group of 70 rats of age 80 days. The mean N content for the first group was 0.983 and the variance was 0.00253; for the second group the corresponding statistics were 1.042 and 0.00224, respectively. Test the hypothesis that the N content does not vary with age. Let $\alpha = 0.01$.

20. A drug manufactured by two different laboratories, A and B, is tested for per cent of purity. The product from A tested in 40 samples shows 94.6

per cent of purity; the product from B tested in 50 samples shows 96.3 per cent. Assume that σ for both is 0.73 per cent. (a) With $\alpha = 0.05$ test the hypothesis that $\mu_B - \mu_A = 1.00$ per cent. (b) What is the size of β if $\mu_B - \mu_A = 1.50$?

21. In a biological experiment the time in seconds elapsed for a reaction after a given stimulus has a variance of 14. A set of 10 individuals shows a mean time interval of 21; another set of 13 shows a mean time interval of 24. Is there an unusual difference in means? Interpret "unusual" as a result corresponding to $P < 0.05$.

In the following exercises assume the distributions of the original variables to be normal so that the t-distribution may be applied where necessary. State clearly what you assume H_0, H_1, or α to be in case they are not specified.

22. In a time and motion study it is found that a certain manual operation averages 36 seconds. A group of 17 workers are given special training and then found to average only 33 seconds with $s = 6$ seconds. Is the argument valid that special training speeds up this operation? Let $\alpha = 0.05$.

23. The numbers of correct answers given by a selected group of 10 persons taking a test were as follows:
$$8, \quad 15, \quad 11, \quad 9, \quad 10, \quad 8, \quad 11, \quad 18, \quad 17, \quad 13.$$
Test the hypothesis that $\mu = 15$ against the alternative that $\mu \neq 15$. Let $\alpha = 0.05$.

24. The scores, x, of 10 students in a psychology test were as follows:
$$61, \quad 70, \quad 85, \quad 81, \quad 76, \quad 79, \quad 94, \quad 43, \quad 67, \quad 74.$$
$$\Sigma x = 730, \quad \Sigma x^2 = 55{,}074.$$
If the mean score of students in general in this test is 70, would you consider this sample an unusual one?

25. The changes in the blood pressure of 18 patients given the same dose of a drug were as follows:
$$6, \ -2, \ 1, \ -1, \ -6, \ 10, \ 3, \ 5, \ -3, \ -4, \ 2, \ 4, \ -2, \ 8, \ 1, \ 3, \ 7, \ 8.$$
Is the assertion that this drug generally increases the blood pressure substantiated?

26. After remedial work in arithmetic a class of 11 pupils showed the following changes in scores on a test.
$$+6, \quad +1, \quad +10, \quad -2, \quad +3, \quad -1, \quad +5, \quad +2, \quad -6, \quad +8, \quad +3.$$
Is the claim justified that this remedial work is beneficial? Let $\alpha = 0.05$.

27. A halibut liver oil product was analyzed by means of 8 random samples. The Vitamin A contents (measured in international units per gram) were as follows:
$$33{,}600, \quad 24{,}900, \quad 24{,}600, \quad 20{,}700, \quad 27{,}000, \quad 25{,}100, \quad 0, \quad 18{,}400.$$
Test the hypothesis that μ for Vitamin A equals at least 27,000. Choose $\alpha = 0.05$.

28. Two laboratory assistants make 10 observations each on the same galvano-meter for the same experiment. The average readings were 61 and 58 with variances of 0.60 and 0.40 respectively. Comment on the difference between the two readings.

29. Fifteen city drug stores showed a mean percentage of profit of 6.37 with a variance of 0.932. Ten suburban drug stores showed a mean percentage of profit of 7.04 with a variance of 1.21. Do these statistics refute the hypothesis that there is no difference in profit between city and suburban stores?

30. Four rats were fed a special ration during the first three months of their lives. The following gains in weight (grams) were noted: 55, 62, 58, 65. Test the hypothesis that $\mu = 65$ against the alternative that $\mu < 65$. Let $\alpha = 0.10$.

31. Ten pupils from one school have a mean I.Q. of 108 and a variance of 60; 17 pupils from another school show a mean of 114 with a variance of 80. Is there a significant difference ($\alpha = 0.05$) between the mean I.Q.'s?

32. The microflavin content (measured in micrograms per 100 grams) of two varieties of turnip greens was determined from samples of 40 turnips each from each variety. The results follow:

$$\text{Variety X:} \quad \Sigma x = 100,000, \quad \Sigma x^2 = 258,800,000;$$
$$\text{Variety Y:} \quad \Sigma y = 92,000, \quad \Sigma y^2 = 219,080,000.$$

Is there evidence of a higher content in variety X than in variety Y?

33. Two machines stamp out washers. During one morning their products were tested by taking random samples from their lots. The outcomes of the measurements made on the two samples are given.

Machine A	Machine B
$\bar{x} = 2.33$	$\bar{y} = 2.61$
$s_x = 0.24$	$s_y = 0.30$
$N_x = 21$	$N_y = 21$

Test the hypothesis, H_0, that $\mu_x = \mu_y$ against the alternative that $\mu_x \neq \mu_y$.

34. Two tinctures of strophanthus were tested by the cat method, each tincture being administered to 7 cats. The mean lethal dose in cubic centimeters of undiluted tincture per kilogram of cat was 0.0168 for tincture A, and 0.0199 for tincture B. The respective standard deviations were 0.00328 and 0.00309. Do the tinctures appear to have significantly different effects?

35. Given the following five pairs of values:

$$\begin{array}{cccccc} x & 5 & 8 & 12 & 6 & 10 \\ y & 2 & 7 & 8 & 7 & 7 \end{array}$$

Using the method of pairing, test the hypothesis that $\mu_d = 0$ against the hypothesis that $\mu_d \neq 0$. Here $d = x - y$. Let $\alpha = 0.05$. What effect would it have on your conclusion if 25 pairs instead of 5 had yielded $\Sigma d = 50$ and $\Sigma d^2 = 180$?

36. Twenty pigs were paired according to equal weight. The two pigs of each pair were then fed different rations for a period of time. The data below show the daily gain in weight in ounces per day. Do the rations produce significantly different mean gains in weight?

Pair	1	2	3	4	5	6	7	8	9	10
Ration X	21	21	19	16	26	19	18	29	22	19
Ration Y	30	25	25	16	29	18	18	19	24	22

37. An attempt is made to compare the effectiveness of two teachers in a certain course. Pairs of pupils are selected who are as nearly alike as possible in previous preparation and in other pertinent attributes. Thus any difference in the test scores of a matched pair will probably not be due to a difference in native ability or to a difference in quality of work done prior to the work to be tested. The data that follow arise from a test given after completion of the course. Test the hypothesis that the teachers are essentially of the same effectiveness.

Pair	1	2	3	4	5	6	7	8	9	10	11	12
						Scores						
Teacher A	41	35	28	39	40	24	26	32	29	41	36	34
Teacher B	36	37	32	38	43	20	22	32	25	42	30	35

38. Systolic blood pressure readings were made on a sample of 15 college men by means of two different types of sphygmomanometers. A reading from each instrument was made on the left arm when the subject was in a sitting position. The readings follow.

Pair	1	2	3	4	5	6	7	8
Sphyg. C (x):	136	138	129	145	158	170	111	125
Sphyg. W (y):	138	145	130	148	166	173	117	120

Pair	9	10	11	12	13	14	15
Sphyg. C (x):	129	144	115	132	141	126	132
Sphyg. W (y):	135	144	120	140	143	132	137

Test the hypothesis that the two instruments do not differ materially in their readings. Let $\alpha = 0.05$.

39. Assume sample sizes, $N = 25$ with $\sigma = 2$. From $\bar{x}$ we are to test the hypothesis, H_0, that $\mu_0 \leq 20$ against the alternatives $\mu_1 = 20.5$, $\mu_2 = 21$, and $\mu_3 = 22$. When $\bar{x} > 20.7$ we reject H_0. (a) What is the value of α? (b) Calculate β for each of the alternatives. (c) Plot the graph of the power function from the values obtained in (b).

40. A certain kind of linen thread is to have a mean breaking strength of 25.8 ounces with a variance of 2.34. When tests are made of random samples of 16 pieces each, a mean breaking strength of 25.0 or lower causes rejection of the lot from which the sample was drawn. (a) What is the size of the Type I error? (b) Compute the size of the Type II error for the alternatives $\mu_1 = 24.5$, $\mu_2 = 25.4$, and $\mu_3 = 25.2$. (c) Sketch the power function.

41. Suppose that hand grenades are timed to explode in five seconds after the catch is released. If you were the manufacturer of these grenades and testing the timing mechanisms periodically, (a) would you use a one-tail or a two-tail test? Why? (b) Would you choose a larger or a smaller Type I error for these tests than you would for testing other nonmilitary timing mechanisms. Why?

42. A manufacturer turns out steel axles with a standard deviation of their diameters equal to 0.50 mm. A buyer desires a sampling plan such that there is a 95 per cent chance of accepting shipments of axles with mean diameters of 10.3 mm. or less, and a 10 per cent chance of accepting shipments with mean diameters of 10.8 or more. What should the sample size for testing be and what is the critical value of $\bar{x}$, that is, the value of the sample mean that separates acceptable shipments from those that are not acceptable?

43. For what alternative hypothesis would the two-tail test illustrated in Figure 8-11c be appropriate?

DISTRIBUTIONS
RELATED TO BINOMIAL
AND NORMAL DISTRIBUTIONS

9

"Probable evidence is essentially distinguished from demonstra-
tive by this, that it admits of Degrees."
BISHOP JOSEPH BUTLER
The Analogy of Religion Natural and Revealed, 1736

9.1. The Distribution of a Proportion

If p is the constant probability for success, then for N independent trials $\mu = Np$ and $\sigma^2 = Npq$. We may interpret p as the proportion of individuals or percentage of them (expressed as a decimal) having a given attribute in a population. The standard deviate

$$z = \frac{x - \mu}{\sigma} . \tag{9.1}$$

If p_1 is the proportion of individuals possessing a given characteristic in a sample of N, then $x = Np_1$ is the *actual* number of them in the sample, and Np is the corresponding *expected* number. Then (9.1) may be written

$$z = \frac{Np_1 - Np}{\sqrt{Npq}} , \tag{9.2}$$

whence

$$z = \frac{p_1 - p}{\sqrt{pq/N}} . \tag{9.3}$$

176

If Np and Nq are not too small, say both exceed 5, the distribution of p_1 about p will be practically normal and the table of normal areas may be used to test hypotheses concerning p. An important thing to note is that the standard deviation of p_1, found in the denominator of (9.3), is given by the formula

$$\sigma_p = \sqrt{\frac{pq}{N}}. \qquad (9.4)$$

In the application of Formula (9.2) (Section 7.10) we found it desirable to apply a correction factor of $\frac{1}{2}$ to the numerator of the fraction. In Formula (9.3) it will generally be desirable to apply a corresponding correction factor of $\pm 1/2N$. For right- and left-tail probabilities the signs to be used are respectively minus and plus.

EXAMPLE 1.   Suppose that a random sample of 1000 births in Graustark showed 53 per cent males and 47 per cent females. Test the hypothesis that the sex ratio is $\frac{1}{2}$ against the alternative that it is higher than $\frac{1}{2}$ for males. Let $\alpha = 0.05$.

Since we seek the probability for a proportion of males as great as 0.53 or greater, the correction factor $1/2N = 1/2000 = 0.0005$ is to be subtracted.

$$p_1 = 0.530, \quad p = 0.500, \quad \text{and} \quad \sigma = \sqrt{\frac{0.500 \times 0.500}{1000}} = 0.0158.$$

Then

$$z = \frac{p_1 - \dfrac{1}{2N} - p}{\sigma_p} \qquad (9.5)$$

$$= \frac{0.530 - 0.0005 - 0.500}{0.0158} = 1.87.$$

Since $P(z > 1.87) = 0.0307$, we conclude that the sex ratio is higher than $\frac{1}{2}$ for males in Graustark. Note that here, as in many problems where N is large, the correction factor is negligibly small.

EXAMPLE 2.   A municipal election is expected to be close with neither of the two contending candidates obtaining more than 55 per cent of the votes cast. How large a pre-election polling sample is required if one is to forecast with a 90 per cent chance of being within 2 per cent of the true percentage of votes to be polled by the winning candidate? Discuss the risks to be made in such a prediction.

Although we do not know the true percentage of votes to be cast for a given candidate, let us, for reasons to be stated later, assume $p = q = \frac{1}{2}$. If we are to have a 90 per cent chance of forecasting within 2 per cent of the correct figure, then our forecast must be within the region from $p = 0.48$ to $p = 0.52$ (Figure 9–1).   Under the standard normal curve the area between $z = \pm 1.645$ constitutes 90 per cent of the total area.   Applying Formula (9.3) we have

$$\pm 1.645 = \frac{\pm 0.02}{\sqrt{\dfrac{0.50 \times 0.50}{N}}}.$$

If, on both sides of the equation, we multiply by the radical, square, and then solve for N, we find that $N = 1689$.   Thus our sample must contain 1689 voters.

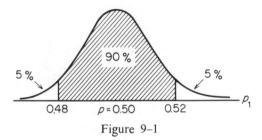

Figure 9–1

If we employ the correction factor, the equation previously used becomes

$$\pm 1.645 = \frac{\pm 0.02 \pm \dfrac{1}{2N}}{\left(\dfrac{(0.50)^2}{N}\right)^{\frac{1}{2}}}.$$

When this equation is cleared of fractions, and simplified, we arrive at the quadratic equation

$$0.0004N^2 - 0.6557N + 0.25 = 0.$$

The larger root of this equation is the useful one here and is 1638.   Thus in this problem, the sample size necessary, computed with greater accuracy, is 1638 instead of 1689.

  Why did we assume that $p = q = \frac{1}{2}$?   The reason for this is related to a familiar question—"For a given perimeter, what must the relative dimensions of a rectangle be in order to have the greatest area?"   The answer, of course, is that the length must equal the width—we must have a square.   Phrased in the language of probability this question becomes the following: "If $p + q = 1$, what values must p and q have in order that pq be a maximum?"   The answer then, is that $p = q = \frac{1}{2}$.   Note that pq is the numerator in the radical above and that the larger this product is, the larger N must be.   Thus this sample of 1689 is surely the largest necessary.

In pre-election polls one encounters a number of risks due to such circumstances as the following: The sample may not be representative of the total voting population; the persons sampled may change their minds about the candidates after the poll has been taken; the majority listed as undecided at the time of the poll may constitute a substantial group favoring one candidate; the persons sampled may not state truthfully how they expect to vote.

9.2. Control Charts for p

The method described in Section 8.4 for controlling the variation in the means of samples may be employed also to study the changes in the

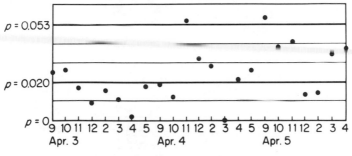

Figure 9–2

fractions of articles found defective in lots of manufactured goods. The three-sigma limits will be $p \pm 3\sqrt{pq/N}$. In practice, p is an estimate, made by averaging previous percentages defective. Samples are taken, usually at regular time intervals, and any erratic behavior or unusual trend in the percentages will show up as the dots on the control chart are plotted.

Inasmuch as the manufacturer is primarily interested in keeping the proportion defective at a minimum, dots lying above the upper control line indicate too large a proportion. Sometimes, however, the fact that dots tend to keep near or below the lower line may indicate an unexpected favorable change in the manufacturing process that may be capitalized upon to reduce the fraction defective.

Figure 9–2 shows a control chart where lack of control appeared at 11 A. M. on April 4 and at 9 A. M. on April 5. The lower limit line is taken as 0, since $p - 3\sqrt{pq/N}$ would be negative.

9.3. The Difference Between Two Proportions

In two samples of N_x and N_y let x and y individuals respectively possess a given attribute, so that $p_1 = x/N_x$ and $p_2 = y/N_y$ are the proportions having this characteristic. If the samples are assumed to be drawn from two independent populations with proportions p_x and p_y, then by Formulas (8.7) and (9.4)

$$\sigma^2_{p_1-p_2} = \sigma^2_{p_1} + \sigma^2_{p_2}$$

$$= \frac{p_x q_x}{N_x} + \frac{p_y q_y}{N_y}. \tag{9.6}$$

If we set up the null hypothesis that x and y stem from the same population, that is, if we assume $p_x = p_y = p$, then we may write

$$\sigma^2_{p_1-p_2} = pq\left(\frac{1}{N_x} + \frac{1}{N_y}\right) \tag{9.7}$$

and employ the standard deviate

$$z = \frac{(p_1 - p_2) - (p_x - p_y)}{\sigma_{p_1-p_2}}$$

so that

$$z = \frac{(p_1 - p_2) - 0}{\left[pq\left(\dfrac{1}{N_x} + \dfrac{1}{N_y}\right)\right]^{\frac{1}{2}}}. \tag{9.8}$$

We test for significance with the aid of the normal curve, provided the values of p and q are not too close to 0 or 1.

Usually p is unknown so we use the following estimates:

$$\hat{p} = \frac{x + y}{N_x + N_y}, \qquad \hat{q} = 1 - \hat{p}; \tag{9.9}$$

$$\hat{\sigma}_{p_1-p_2} = \left[\hat{p}\hat{q}\left(\frac{1}{N_x} + \frac{1}{N_y}\right)\right]^{\frac{1}{2}}. \tag{9.10}$$

The correction factor for continuity, for reasons not easily shown here, comes out equal to

$$\frac{N_x + N_y}{2N_x N_y}. \tag{9.11}$$

For right- and left-tail probabilities the signs to be used with it are, respectively, minus and plus.

EXAMPLE 1.   In an "attitude" test, 45 out of 113 persons of Race R and 112 persons out of 381 persons of Race S answered "yes" to a certain question. Do these races differ in their attitudes on this question? We wish to test the hypothesis that $p_x - p_y = 0$ against the alternative that $p_x - p_y \neq 0$. Choose $\alpha = 0.05$.

$$p_1 = \frac{45}{113} = 0.392, \qquad p_2 = \frac{112}{381} = 0.294;$$

$$\hat{p} = \frac{45 + 112}{113 + 381} = 0.318, \qquad \hat{q} = 0.682;$$

$$\hat{\sigma}_{p_1 - p_2} = \left[0.318 \times 0.682 \left(\frac{1}{113} + \frac{1}{381} \right) \right]^{\frac{1}{2}} = 0.0500.$$

The correction factor is $\dfrac{113 + 381}{2(113 \times 381)} = 0.00574$ or approximately 0.006.

$$z = \frac{(0.392 - 0.294) - 0.006}{0.0500} = 1.84.$$

For a two-tail probability of 0.05, z would have to equal 1.96. Since $z < 1.96$, the hypothesis is not refuted. It is interesting to note that without the correction factor z would equal 1.96. Unless one adheres strictly to a Type I error of 0.05, there might well be grounds for suspecting racial differences to influence the attitudes in question.

9.4. The Poisson Distribution

In applying the normal curve approximation to the calculation of binomial probabilities we assume that p is not very small. If the probability, p, for an event to happen is excessively small, the approximation is poor and a better formula must be sought. In this case, the event may be termed rare. Examples of rare events are drawing the ace of diamonds from a pack of cards, throwing a double six with a pair of dice, or contracting a case of scurvy in Boston. The following theorem, the proof of which cannot be given here, becomes applicable.

In a binomial distribution, if N becomes infinite as p approaches 0 as a limit, in such a manner that Np remains constant (equal to μ), then the distribution approaches as a limit the Poisson distribution

$$f(x) = \frac{e^{-\mu} \mu^x}{x!} . \qquad (9.12)$$

Thus if N is large and p is very small, say $\mu = Np < 5$, an excellent approximation to the probability for exactly x successes in N trials is given by Formula (9.12). The expression on the right of (9.12) is called the *Poisson exponential function.* The probabilities for exactly 0 occurrences, 1 occurrence, 2 occurrences, ..., N occurrences, are given by the respective terms of the series

$$\sum_{x=0}^{N} \frac{e^{-\mu}\mu^x}{x!} = e^{-\mu}\left(1 + \frac{\mu}{1!} + \frac{\mu^2}{2!} + \frac{\mu^3}{3!} + \cdots + \frac{\mu^N}{N!}\right). \qquad (9.13)$$

This, like the binomial, is a discrete distribution, for x assumes only non-negative integral values.

It may be proved that the expected value or population mean

$$\mu = Np, \qquad (9.14)$$

and that the *variance* is equal to it:

$$\sigma^2 = \mu = Np. \qquad (9.15)$$

The distribution is skew for small values of μ but tends to become more symmetrical with increasing μ, as illustrated in Figure 9–3.

It is not always clear when the Poisson distribution should be used rather than the normal to approximate to the binomial. Since the shape of the distribution is determined by the parameter, μ, and is nearly symmetrical when $\mu = 5$ (Figure 9–3), a fairly safe rule is to use the Poisson approximation whenever $\mu < 5$. The binomial distribution is always accurate and in cases of doubt it may be employed.

The values of the individual probabilities given by (9.12) are tabulated in the Biometrika Tables (Reference 25) and in the Fisher and Yates Tables (Reference 20). These probabilities and the cumulated terms beginning with an arbitrary value of x

$$\sum_{x=k}^{N} f(x) = \sum_{x=k}^{N} \frac{e^{-\mu}\mu^x}{x!}$$

are also found in Molina's Tables (Reference 23), which are especially useful.

EXAMPLE 1.   The mortality rate for a certain disease is 7 per 1000. What is the probability for just 5 deaths from this disease in a group of 400?

$$\mu = Np$$

$$= 400 \times 0.007 = 2.8.$$

From Molina's Tables, Part I, page 4, (here $a = \mu = 2.8$, $c = x = 5$), we find that

$$f(5) = \frac{e^{-2.8}(2.8)^5}{5!} = 0.0872.$$

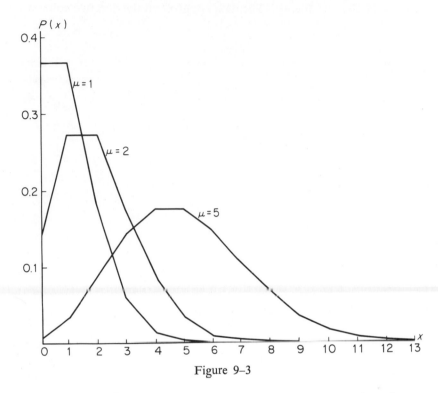

Figure 9-3

EXAMPLE 2.   As a rule $\frac{1}{2}$ per cent of certain manufactured products are defective. What is the probability that 1000 of them will have 10 or more defective?

$$\mu = 1000 \times 0.005 = 5;$$

$$P = \sum_{x=10}^{1000} \frac{e^{-5}5^x}{x!}$$

$$= e^{-5}\left(\frac{5^{10}}{10!} + \frac{5^{11}}{11!} + \frac{5^{12}}{12!} + \cdots\right).$$

From Molina's Tables for cumulated terms, Part II, page 6, we find for

$$a = \mu = 5, \; c = x = 10,$$

that $P = 0.0318$, hence the occurrence of 10 defective products in 1000 might well be considered unusual.

EXAMPLE 3.   Von Bortkiewicz's classic example (1898) used the frequency distribution of the number of deaths from horse kicks per army corps per year in the Prussian army for 20 years, 1875–1894, to illustrate the agreement between actuality and theory. The data are given in the first two columns of Table 9–1.

Table 9–1

No. of deaths x	Actual frequency f	fx	fx^2	Theoretical frequency
0	109	0	0	108.7
1	65	65	65	66.3
2	22	44	88	20.2
3	3	9	27	4.1
4	1	4	16	0.7
		122	196	

Number of deaths: 122
Number of army corps: 10
Number of years: 20

The mean number of deaths per army corps per year

$$\bar{x} = \frac{122}{200} = 0.61.$$

The variance

$$s^2 = \frac{196}{200} - (0.61)^2 = 0.61.$$

Here the agreement between $\bar{x}$ and s^2 is remarkably close and constitutes strong evidence that the distribution is Poissonian in character. Letting $\mu = 0.61$ we obtain the theoretical frequencies by multiplying by 200 the first five terms of (9.13), with $\mu = 0.61$;

$$200e^{-0.61}\left[1 + \frac{0.61}{1!} + \frac{(0.61)^2}{2!} + \frac{(0.61)^3}{3!} + \frac{(0.61)^4}{4!}\right]$$

$$= 108.7 + 66.3 + 20.2 + 4.1 + 0.7.$$

The agreement between the actual and theoretical frequencies is excellent.

9.5. The Multinomial Distribution

We recall that the binomial distribution is discrete with probabilities given by the successive terms of the expansion of $(q + p)^N$. It is concerned with two mutually exclusive categories, occurrence and nonoccurrence of an event. In certain problems the labor of computing a desired sum of binomial probabilities is avoided by approximating to it by means of the area under a continuous curve, that of the normal distribution.

A similar situation arises when more than two mutually exclusive categories exist. For example, in tossing six pennies any number of heads from 0 to 6 may occur, hence there are seven mutually exclusive categories. It may be proved that if $p_1, p_2, \ldots, p_k$ are the respective probabilities of obtaining the mutually exclusive values $x_1, x_2, \ldots, x_k$, then the probability P, of obtaining in N trials, x_1 just n_1 times, x_2 just n_2 times, $\ldots, x_k$ just n_k times, where $\sum_{i=1}^{k} n_i = N$, is given by the formula

$$P = \frac{N!}{n_1!\, n_2! \cdots n_k!}\, p_1^{n_1} p_2^{n_2} \cdots p_k^{n_k} \tag{9.16}$$

which is a general term derived from the multinomial expansion

$$(p_1 + p_2 + \cdots + p_k)^N. \tag{9.17}$$

Formula (9.16) defines the *multinomial probability function*. Its derivation is straightforward and resembles that of Section 6.7.

The particular sequence of events

$$\overbrace{x_1 x_1 \ldots x_1}^{n_1} \overbrace{x_2 x_2 \ldots x_2}^{n_2} \cdots \overbrace{x_k x_k \ldots x_k}^{n_k},$$

where x_i occurs n_i times in succession, has a probability

$$p_1^{n_1} p_2^{n_2} \cdots p_k^{n_k} \tag{9.18}$$

of occurring, since all N trials are independent of one another. Any other sequence has the same probability associated with it and there are

$$\frac{N!}{n_1!\, n_2! \cdots n_k!} \tag{9.19}$$

different sequences possible [see Formula (6.3.)], hence the probability for any one of these sequences is given as the product of (9.18) and (9.19) which is the multinomial formula (9.16).

Example. A community consists of 50 per cent English, 30 per cent Irish, and 20 per cent Scottish persons. If a sample of six individuals is selected at random, what is the probability that two are English, three are Irish, and one is a Scot?

The probabilities for persons of these three races are respectively, 0.50, 0.30, and 0.20. Then

$$P = \frac{6!}{2!\, 3!\, 1!}\, (0.5)^2 (0.3)^3 (0.2) = 0.081.$$

Multinomial probabilities usually require extensive computation unless one makes use of approximations that are obtained with the aid of the *Chi-square function*. The latter is studied in Chapter 11.

EXERCISES

1. In crossing certain varieties of peas it was expected that 25 per cent of the seeds would be green. However it was found that of 3675 seeds 26.3 per cent were green. Could this be the result of chance?

2. It is claimed that 20 per cent of the voters in a certain community are independent voters. A representative sample of 236 voters are polled, and of these, 40 state that they are independent. On the basis of this sample is the claim justified?

3. On an average 32 per cent of the persons afflicted with a certain malady die. If under a new treatment 24 per cent out of 200 die, might you say that the new treatment is effective? Let $\alpha = 0.04$.

4. The failures in a qualifying examination have been found to average 25 per cent. (a) What can you say about a group of 156 pupils from School X who show 26 failures? (b) How large a sample is necessary in order that one may be 90 per cent sure that the per cent of failures does not exceed 30 per cent? You may omit the correction factor in solving (b).

5. A bag contains a very large number of black and white beads all of uniform size and weight. There are twice as many white beads as black. If 24 beads are selected at random what is the probability that 20 or more of them are white?

6. In the manufacture of a type of delicate filament the per cent of defectives is found to average 12. It is decided to adopt a modified method of manufacture if and only if a random sample of 300 filaments shows the per cent of defectives equal to 9 or less. (a) What is the probability of adopting the new method if the mean per cent defective remains equal to 12? (b) What is the size of the type II error if the new method actually reduces the per cent defective to 8? (c) After the new method had been tried a sample of 300 showed 25 filaments defective. Should the new process be adopted?

7. If 10 per cent of the mice injected with a certain serum die, how large a group of mice must be tested in order that there be a probability of 0.02 that 5 per cent or fewer die? Obtain the answer in two ways; (a) without and (b) with the use of the correction factor.

8. An insecticide usually kills 80 per cent of the insects sprayed. How large a sample of them should be taken in order that we may be 95 per cent sure that the per cent killed differs from the expected per cent by less than 10? Omit the correction factor.

9. Extensive experience in the manufacture of a certain type of galvanized hardware showed the percentage of surface defects to average 1.30. Construct a control chart for p with 3σ limits for the following results, where $N = 580$.

Lot no.:	1	2	3	4	5	6	7
No. defective:	9	3	7	9	4	8	8

10. In an opinion poll 40 out of 160 women and 80 out of 240 men answered "yes" to a certain question. Test the hypothesis that the two sexes do not differ significantly on this matter, against the alternative that they do. Let $\alpha = 0.05$.

11. In a competitive examination, 45 pupils out of 600 in School X and 40 pupils out of 800 in School Y received scores of 90 or better. May one conclude, in general, that School X has more superior students than School Y?

12. In a certain college, the senior class consisted one year of 231 men and 193 women. From this class 19 men and 20 women were elected to Phi Beta Kappa. Does this indicate that, generally, a larger fraction of women are given this honor than men?

13. A group of 200 students was given an examination. One hundred and thirty students answered question 2 correctly and 148 answered question 3 correctly. Test the hypothesis that these two questions are of equal difficulty, against the alternative that question 3 is easier. Let $\alpha = 0.10$.

14. The following data are given in the form of a 2×2 table. Let p_1 be the proportion of recoveries among those inoculated and p_2 the proportion among those not inoculated. Test the hypothesis that inoculation does not affect recovery, with a significance level of 2 per cent. State your hypothesis and the alternative in the form of equations in the p's. Why is it safe to omit the correction factor here?

	Recoveries	Deaths	Total
Inoculated	89	53	142
Not inoculated	32	61	93
Total	121	114	235

15. Do the data shown below indicate any real difference between Methods A and B used in a manufacturing process?

	Defective	Nondefective
Method A	41	20
Method B	16	28

16. The number of vacancies on the Supreme Court Bench filled by the presidents of the United States up to 1932 were as follows:

No. vacancies	Frequencies
0	59
1	27
2	9
3	1
Over 3	0

Show that the distribution is of the Poisson type by computing the theoretical frequencies.

17. The author's records of freshman classes averaging 25 students per class show that the number of A+ grades given per class were as follows:

No. grades	Frequency
0	24
1	10
2	1
3 or more	0

Show that the distribution is of the Poisson form.

18. As a rule one per cent of certain units purchased by a manufacturer are unfit for use. What is the probability (a) that 12 or more in a lot of 500 will be unfit? (b) that fewer than 4 in a lot of 400 will be unfit?

19. About 9800 pairs of twins occurred among every 1,000,000 confinements between 1940 and 1944. If a maternity hospital had 256 confinements in 1942 with 7 pairs of twins, would you say that this was a normal year for twins in this hospital?

20. A Sales-by-Mail company receives orders, on an average, from 2 per cent of the letters sent out. A new type of sales appeal is tried in 1000 letters sent out, and 30 orders are received. Is the new type more effective than the old?

21. A factory produces items which are usually 5 per cent defective. The process of manufacture is modified and then tested for possible improvement. A random sample of 200 shows only 6 defective. Would you say ($\alpha = 0.05$) that the modification reduces the number of defectives? Compare the probabilities obtained by using both the Poisson and binomial distributions.

22. If, on an average, 2 lost articles per day are reported on a certain railroad train, what is the probability that more than 4 are reported on a given day?

23. As the result of a psychological test, candidates are rated as (1) superior; (2) average; (3) inferior. In general the per cents falling into these categories are 20, 50, and 30 respectively. If the test is given to 8 persons, what is the probability that 2 are rated superior, 3 average, and 3 inferior?

24. A package of candy contains 20 wafers of 5 different flavors and colors. The mixing apparatus in the candy factory uses wafers in the following proportions but mixes them randomly:

	Per cent
White (peppermint)	30
Brown (chocolate)	20
Yellow (lemon)	35
Green (lime)	10
Black (licorice)	5

If a boy purchases a package of these wafers, what is the chance that he gets a package mixed exactly in these proportions? Do not attempt to evaluate your result.

25. A die has 3 faces colored red, 2 colored white, and 1 colored blue. If this die is thrown 9 times what is the probability that each color appears 3 times? Compute to 2 decimal places.

26. In a research on "flying bombs" fallen in London during World War II, an area of 144 square kilometers in the south of London was divided into 576 squares of $\frac{1}{4}$ square kilometer each. The results are shown in the table that follows. Show that the distribution is of the Poisson type (a) by comparing the mean with the variance, and (b) by comparing the theoretical frequencies with the actual.

No. bombs fallen	No. observed squares
0	229
1	211
2	93
3	35
4	7
5 and over	1

[Data used by permission of R. D. Clarke, *Journal of the Institute of Actuaries*, London. Vol. 72, 1946, p. 481.]

ESTIMATION

10

10.1. Introduction

A professor of statistics who wished to introduce the material of this chapter used to begin by inquiring of his class, "How old do you think I am?" Before receiving their replies he assured the students that no reprisals would ensue if their guesses exceeded his age. The replies were invariably in the form "52 years;" "49 years;" "58 years;" and so on. Each guess was always an "all or nothing" guess, that is to say, each answer was always given as an exact number of years and was either right or wrong. If it was wrong there was no way of knowing how much it was in error. There was no way of calculating, even roughly, the probability of a reply being the correct one. Such "all or nothing" estimates are called *point estimates* in statistics.

The professor next asked his students if there wasn't some other way of guessing his age so that limits could be placed on his probable age. Usually this type of response followed. "You are between 50 and 55;" "You are between 45 and 55;" "You are between 40 and 60;" and so forth. Now the estimates of his age took the form of intervals. In the case of the teacher and the guess of 40 to 60 years, the answer was felt to be almost surely correct and the students were willing to place heavy odds on a bet that this was a correct statement. As for the first guess, 50 to 55, there was a distinct drop in the degree of confidence in the estimate—appearances are deceptive and young people are not usually expert in estimating the ages of older people. This second type of estimate is called an *interval estimate*. In statistics we shall find that it has the advantage that the interval can be adjusted, that is, widened or narrowed, with a corresponding increase or reduction in confidence of being correct. For this reason we shall speak of the *confidence interval* as a method of statistical estimation. Both types of estimates are in common use and each has its special advantages.

10.2. The Point Estimate

Given a sample mean $\bar{x}$, what is the best estimate one can make of the mean μ, of the population from which the sample is drawn? In the lack of other information our intuition leads us to say that $\bar{x}$ is the best estimator of μ. Furthermore, the larger the sample size is, the greater our faith in the accuracy of this estimate. By the methods of mathematical statistics one can prove that our intuition is correct, so that we shall state, without proof

THEOREM 10.1. *If $\bar{x}$ is the mean of a random sample of N drawn from any population with mean μ, then an unbiased estimator of μ is $\bar{x}$.* In symbols,

$$\hat{\mu} = \bar{x}. \tag{10.1}$$

The word *estimator* refers to the function, *estimate* to a particular value of it. An estimator of a parameter such as μ is designated by placing a circumflex accent above the symbol for the parameter. We have introduced the word *unbiased*. Briefly, this means that if repeated samplings of the same population took place, the mean of the sample means would tend toward the population mean. Some estimates have a tendency to be biased,—to be either too large or too small. The sample mean is unbiased. There are estimates other than unbiased ones and these are mentioned briefly in Section 10.11.

10.3. The Unbiased Estimate of p

Let p be the proportion of individuals having a given characteristic in a population and let a random sample of N be drawn from this population. If x individuals out of N are found to possess the given characteristic, then x/N would seem to be the best estimate that we could make of the population proportion p. This intuition can be proved to be correct, so we state, without proof

THEOREM 10.2. *If x is the number of individuals possessing a given attribute in a random sample of N drawn from a population in which p is the proportion of individuals possessing this attribute, then x/N is an unbiased estimator of p.* In symbols,

$$\hat{p} = \frac{x}{N}. \tag{10.2}$$

10.4. Estimates of σ^2

Following the discussion of the preceding two sections, one might guess that an unbiased estimator of the population variance σ^2, is s^2, where $s^2 = \dfrac{1}{N}\Sigma(x - \bar{x})^2$. If N is large this guess is close to the truth, but if N is small, then a serious error might be committed. The explanation is not difficult to follow. By Theorem 2.6 of Section 2.11, the sum of the squares of the deviations of the sample values from their own mean $\bar{x}$, will be smaller than the sum of the squares of the deviations from the true mean, μ. Thus s^2 tends to be smaller than σ^2. If N is large, $\bar{x}$ is probably very close to μ so that the bias is negligibly small.

We state without formal proof

THEOREM 10.3. *An unbiased estimator of the population variance* σ^2, *is given by the formula*

$$\hat{\sigma}^2 = \frac{1}{N-1}\Sigma(x - \bar{x})^2 \tag{10.3}$$

$$= \frac{Ns^2}{N-1}. \tag{10.4}$$

Since s^2 tends to be too small, this replacement of N by $N-1$ in the denominator of the formula for s^2 offsets this bias. Because of the foregoing facts, some statisticians prefer to define the sample variance as in (10.3). In consulting other books and papers in statistics, the reader should take care to note what definition of s^2 is used.

We may also prove by (10.4) that an unbiased estimator of the variance of the mean

$$\hat{\sigma}_{\bar{x}^2} = \frac{\hat{\sigma}_{x}{}^2}{N} \tag{10.5}$$

$$= \frac{s^2}{N-1}. \tag{10.6}$$

If N is large, the replacement by $N-1$ is a negligible improvement. If N is small, this replacement is important. Small-sample theory and large-sample theory are generally different.

The symbol, $\hat{\sigma}^2$, is not to be construed as $(\hat{\sigma})^2$ because it does not follow from (10.3) that an unbiased estimator of the standard deviation,

$$\hat{\sigma} = \left[\frac{1}{N-1}\Sigma(x - \bar{x})^2\right]^{\frac{1}{2}},$$

although it is frequently used as such. Unfortunately, the reason for this seeming paradox cannot be stated here. Suffice it to say that the frequency functions of s^2 and s are unlike.

If two samples of N_x and N_y are assumed to be drawn from the same population and yield variances s_x^2 and s_y^2 respectively, it can also be proved that a useful unbiased estimator of the population variance, σ^2, is obtained by pooling the two sample variances. The formula is in fact the following:

$$\hat{\sigma}^2 = \frac{N_x s_x^2 + N_y s_y^2}{N_x + N_y - 2}. \tag{10.7}$$

This result can be generalized for k samples, so that

$$\hat{\sigma}^2 = \frac{\sum\limits_{i=1}^{k} N_i s_i^2}{\sum\limits_{i=1}^{k} N_i - k}. \tag{10.8}$$

10.5. Confidence Limits for μ

Suppose that a random sample of N has been drawn from a normal population with unknown mean, μ and unknown variance, σ^2. From the information, N, $\bar{x}$, and s, contained in the sample, we wish to estimate the value of μ. If we made a point estimate of this value we should choose $\bar{x}$, but we should have no way of knowing how good this estimate might be. Instead, we shall calculate a *confidence interval* within which, it will be stated, μ lies. A measurable degree of confidence will be associated with this statement. The reason that we cannot make a probability statement about μ is that μ is not a chance variable, it is a constant, fixed, but unknown. Probability statements are always associated with chance or random variables. (See Section 1.5.)

To illustrate the method let us suppose that a random sample of 26 electric light bulbs shows a mean length of life of 1000 hours and a standard deviation of 200 hours. We seek the 95 per cent confidence limits for μ, the mean life of the lot of bulbs from which this sample was drawn.

If the lives of bulbs are assumed to be normally distributed, we may employ the Student-Fisher t-variable defined in Formula (8.11). From Table E, when $n = 25$ and $P = 0.025$, $t = +2.06$ (Figure 10–1). Thus we may determine 95 per cent confidence limits for μ as follows:

For $t = \pm 2.06$ and the given data, Formula (8.11) becomes

$$\pm 2.06 = \frac{(1000 - \mu)\sqrt{25}}{200}.$$

Solving for μ we have for the plus sign $\mu = 918$ and for the minus sign $\mu = 1082$. We now make the following statement: $918 < \mu < 1082$. This statement may be wrong, because the sample might be unusual in its mean or in its standard deviation or in both. In such a case, the value of t would deviate abnormally from its mean, $\mu_t = 0$, and the limits for μ based upon this value would be incorrect. On the other hand, if the sample exhibits only the usual fluctuations (say those occurring 95 per cent of the time), then

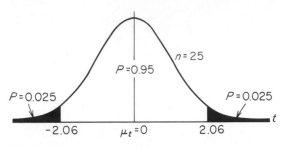

Figure 10–1

the double inequality above will be correct. We now assert the following (see Figure 10–1):

(1) There is a probability of 0.95 that $-2.06 < t < 2.06$.

(2) For a sample of 26, there is a probability of 0.95 that

$$-2.06 < \frac{(\bar{x} - \mu)\sqrt{25}}{s} < 2.06$$

or that

(3) $-0.412s < \bar{x} - \mu < 0.412s$.

This implies that 95 per cent of the time

(4) $\mu - 0.412s < \bar{x} < \mu + 0.412s$.

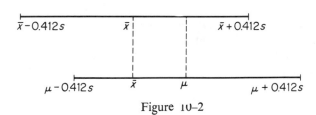

Figure 10–2

If a given $\bar{x}$ does lie in such an interval (lower portion of Figure 10–2), then by virtue of (4) μ lies in the interval

$$\bar{x} - 0.412s \quad \text{to} \quad x + 0.412s,$$

(upper portion of Figure 10–2). But if a given $\bar{x}$ does *not* lie in this interval (Figure 10–3) and this happens in 5 per cent of the samples, then μ does not lie in the interval defined above.

In the illustrative problem $N = 26$, $\bar{x} = 1000$, and $s = 200$; hence μ lies in the interval

$$1000 - (0.412 \times 200) \quad \text{to} \quad 1000 + (0.412 \times 200)$$

that is, $918 < \mu < 1082$.

In words, then, we state that the mean length of life of the entire lot of bulbs lies between 918 and 1082 hours. Statements such as this will be correct 95 per cent of the time and incorrect 5 per cent of the time. This means that,

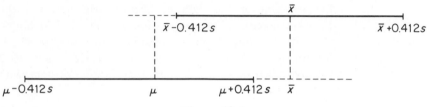

Figure 10–3

under the same conditions, repeated samplings of 26, leading to repeated inequality statements, would produce 95 per cent correct statements because 95 per cent of the time the sample means would lie within 0.412s of the population mean. Of course, any confidence interval other than one for 95 per cent may be used. Clearly, the higher the confidence level, the wider the confidence interval, and the lower the confidence level, the narrower the interval.

A general formula for confidence limits for μ is

$$\bar{x} - \frac{s}{\sqrt{N-1}} t_{\alpha/2} < \mu < \bar{x} + \frac{s}{\sqrt{N-1}} t_{\alpha/2}, \tag{10.9}$$

where $t_{\alpha/2}$ is the right-tail value of t corresponding to $100(1 - \alpha)$ per cent of confidence.

10.6. Confidence Limits for $\mu_x - \mu_y$

Given two normal populations from which samples of N_x and N_y have been drawn, with means $\bar{x}$ and $\bar{y}$ and variances s_x^2 and s_y^2 respectively, confidence limits can be computed for $\mu_x - \mu_y$ in the following manner. In Formula (10.9) above, replace $\bar{x}$ by $\bar{x} - \bar{y}$, $\dfrac{s}{\sqrt{N-1}}$ by the denominator of Formula (8.13) and μ by $\mu_x - \mu_y$.

10.7. Confidence Limits for μ_d

Formula (8.14) enables one to calculate desired confidence limits for μ_d, the population mean of the differences between paired observations.

10.8. Confidence Intervals and Tests of Hypotheses

One advantage derived from the use of confidence intervals, is, that in addition to making estimates, they can be used to test hypotheses concerning unknown parameters. A confidence interval can be treated as an acceptance region for the hypothesis.

EXAMPLE 1. Let us refer to the data of Section 10.5 and test the hypothesis that the mean length of life of the lot of bulbs is 950 hours, against the alternative that it is not. Using $\alpha = 0.05$ to correspond to 95 per cent limits, we note that $\mu = 950$ lies within the interval $918 < \mu < 1082$ already found. The hypothetical value, 950, is, therefore, consistent with this interval estimate based on the sample data.

EXAMPLE 2. (A one-sided confidence interval). A random sample of 17 pieces of thread is selected from a shipment and has a mean breaking strength of 17.4 ounces and a standard deviation of 2.13 ounces. With $\alpha = 0.05$, test the hypothesis that μ for the shipment is at least 18 ounces.
Here $N - 1 = 16$, $\bar{x} = 17.4$, $s = 2.13$, and $t = -1.746$.

Therefore
$$-1.746 = \frac{(17.4 - \mu)\sqrt{16}}{2.13}$$

whence $\mu = 18.3$. Thus the upper limit for μ is estimated to be 18.3 ounces. Our hypothesis that μ is at least 18 is consistent with this estimate.

10.9. Confidence Limits for p

Example. A pre-election poll is taken in a town by means of a carefully selected sample of 50 voters. Of these 40 per cent declare that they will vote for Bill Jones to be Mayor. Find the 90 per cent confidence interval for the proportion in the town who will vote for Jones.
From (9.5)

$$\pm z = \frac{p_1 \mp \dfrac{1}{2N} - p}{\left[\dfrac{p(1 - p)}{N}\right]^{1/2}} ; \qquad (10.10)$$

Hence
$$\pm 1.645 = \frac{0.4 \mp 0.01 - p}{\left[\dfrac{p(1 - p)}{50}\right]^{1/2}} \qquad (10.11)$$

where the upper signs are used for the lower confidence limit and the lower signs are used for the upper one. By squaring and simplifying (10.11) we obtain the quadratic equations

$$1.0541 p^2 - 0.8341 p + 0.1521 = 0, \quad \text{(upper signs)} \qquad \textbf{(10.12)}$$

$$1.0541 p^2 - 0.8741 p + 0.1681 = 0, \quad \text{(lower signs)} \qquad \textbf{(10.13)}$$

with roots 0.505 and 0.285 for the first equation and 0.526 and 0.302 for the second. Since Equation (10.12) is not equivalent to (10.11) with the upper sign, we find that only the root, 0.285, satisfies the latter. Similarly, only the root, 0.526, satisfies (10.11) with the lower sign. Then the 90 per cent confidence interval is defined by the double inequality $0.285 < p < 0.526$.

A simplification of the computational labor can be obtained by following this working rule:

When N is large or when $0.3 < p_1 < 0.7$, a good approximation to the confidence limits may be derived by replacing $p(1 - p)$ in the denominator of (10.10) by $p_1(1 - p_1)$.

If we use this approximation we have

$$\pm 1.645 = \frac{0.40 \mp 0.01 - p}{\left[\dfrac{0.40 \times 0.60}{50} \right]^{\frac{1}{2}}},$$

whence for the upper signs, $p = 0.276$, and for the lower signs $p = 0.524$. Then $0.276 < p < 0.524$.

It is readily seen that there is no practical difference between the answers obtained by these two methods, hence the shorter one is recommended when the appropriate conditions are met. Observe also that, even if we have unbounded faith in the representativeness of the sample and the veracity of the sampled voters, a prediction that the vote for Jones will be between 0.28 and 0.52 of the total vote cast, will not enable us to decide, with 90 per cent confidence, the outcome of the election. If, however, we reduce our confidence level to 80 per cent ($z = \pm 1.282$), then we may state that $0.30 < p < 0.50$. Thus our prediction is that Jones will be defeated, but our confidence in this prediction is at a lower level.

The calculations of this section can be avoided by referring to charts showing "Confidence Belts for p." These appear in a paper by C. J. Clopper and E. S. Pearson, "The use of confidence or fiducial limits illustrated in the case of the binomial," *Biometrika*, Volume 26 (1934), pages 410 and 411. Table H in this text reproduces such a chart for a confidence coefficient of 0.95. From this chart one can find readily the 95 per cent confidence interval for the Jones vote to be $0.27 < p < 0.55$. One may also find confidence limits for p from the tables of Mainland, Herrera, and Sutcliffe (Reference 22).

10.10. Confidence Limits for σ^2

Given a sample of N with variance s^2, one can establish confidence limits for the population variance σ^2. This is done by means of the Chi-square distribution and will be discussed in the next chapter (Section 11.3).

10.11. Properties of Point Estimates

In Section 10.2 we introduced the concept of an unbiased estimate of a parameter. More precisely we may state the following definition:

The statistic $\hat{\theta}$ is an unbiased estimator of the parameter θ if $E(\hat{\theta}) = \theta$

Here $E(\hat{\theta})$ is the expected value or theoretical mean of $\hat{\theta}$. (See Section 5.7.) For example, the distribution of $\bar{x}$ from a normal population is known to be normal with mean $\mu_{\bar{x}} = \mu_x$. Thus $\bar{x}$ is an unbiased estimator of μ_x. Usually, but not always, $\bar{x}$ lies fairly near μ because the mean of the $\bar{x}$'s approaches μ as a limiting value.

An unbiased estimator, $\hat{\theta}$, may occasionally deviate considerably from θ, therefore we take into consideration its variability with respect to θ measured in terms of the variance. A desirable property, then, of an estimator, $\hat{\theta}$, is that its variance should be as small as possible; in other words, it should be *efficient*.

The statistic $\hat{\theta}$ is an efficient estimator of the parameter θ if the variance of $\hat{\theta}$ is not greater than the variance of any other estimator of θ.

The concept of the *relative efficiency* of two estimates is an important one, especially in many practical applications.

The efficiency of an unbiased estimate $\hat{\theta}'$ relative to some other unbiased estimate $\hat{\theta}$ is defined as the ratio of $\sigma_{\hat{\theta}}^2$ to $\sigma_{\hat{\theta}'}^2$.

For example, the mean $\bar{x}$ and the median x_m are both measures of central tendency. To compare their efficiencies as estimates of the population mean μ, we divide $\sigma_{\bar{x}}^2$ by $\sigma_{x_m}^2$. In the case of a normal population and a sample of 10 it has been shown that this quotient is about 0.72. Thus the efficiency of the median x_m, is 0.72 relative to that of the mean $\bar{x}$, for samples of 10. This means that the accuracy of the estimate of μ from the median of a sample of 100 is approximately the same as that from the mean of a sample of 72. Thus fewer observations are necessary for efficient estimators than for less efficient ones.

It can be proved that the sample mean $\bar{x}$, is an unbiased and efficient estimator of μ, and that the sample variance

$$s^2 = \frac{1}{N} \sum_{i=1}^{N} (x_i - \bar{x})^2$$

is a biased estimator of σ^2.

However, $$\frac{N}{N-1} s^2 \quad \text{or} \quad \frac{1}{N-1} \sum_{i=1}^{N} (x_i - \bar{x})^2$$

is unbiased. The latter is an efficient but not exactly the "most-efficient" estimator of σ^2, for its efficiency can be shown to be $N - 1/N$ which approaches 1 (100 per cent) as N becomes larger and larger. The estimator $\frac{1}{N} \sum (x - \mu)^2$ has a smaller variance than $\frac{1}{N-1} \sum (x - \bar{x})^2$ but it is biased.

EXERCISES

1. In Exercise 3.6 the mean brain-weight of the Swedish males is 1400.5 grams. In Exercise 3.12 the mean thickness of the washers is 0.1000 inches. What is an unbiased estimate of (a) the mean brain-weight of the population of Swedish males? (b) the mean thickness of the lot of washers of which those tabulated are a sample? (c) Which of these two estimates is more likely to be more accurate? Why?

2. The reading rates (story-book reading) of 10 junior high school pupils before and after remedial work showed means of 218 and 257 words per minute respectively. Make a point estimate of the mean gain in rate to be expected from similar work in the future.

3. Assume that random samples of 80 men voters and 60 women voters in a village showed 50 men and 40 women registered as Republicans. Make a point estimate of the percentage in the village of (a) male Republicans; (b) female Republicans; (c) Republicans.

4. The mean mileage obtained from a sample of 17 automobile tires of a taxi-cab company was 28,000 with $s = 1800$. Find the 90 per cent confidence interval for the mean life of tires used by this company.

5. Twenty-six typical army recruits have a mean pulse rate of 71.2 beats per minute with a variance of 83.4. Find the 95 per cent confidence limits for the mean pulse rate of recruits in general. What does your answer mean?

6, If the mean age at death of 64 men engaged in a somewhat hazardous occupation is 52.4 years with a standard deviation of 10.2 years, what are the 98 per cent confidence limits for the mean age of all men so engaged?

7. A random sample of 15 capsules of a halibut liver oil product was tested for Vitamin A content. The mean amount (in 1000 international units per gram) was found to be 23.4 with a standard deviation of 8.50. Find the 90 per cent confidence interval for μ.

8. A sample of 17 shows a product to have a mean per cent of purity of 91.3 with $s = 1.43$. What are the 95 per cent confidence limits for the mean per cent of purity of the population?

9. A special aptitude test was given to 26 law school freshmen. The results showed a mean score of 82.0 and a variance of 49.00. Set up the 90 per cent confidence interval for the mean score of all law school freshmen.

10. After a fixed dose of a certain drug had been administered to 17 selected students, it was found that their pulse rates increased by the following amounts:

13, 15, 14, 10, 8, 12, 15, 9, 18, 10, 16, 11, 7, 15, 13, 16, 11.

In general, pulse rates of students increase by 10 for the same dose. Is the mean increase in this group significantly greater than the mean for all students in general? Solve by two methods, one of which requires finding 95 per cent confidence limits for μ.

11. In Exercise 8.22 find the one-sided 95 per cent confidence interval for the mean time required when *all* workers are given the special training. Does the given population mean, 36 seconds, lie above the upper limit of this interval? What does your answer indicate?

12. In Exercise 8.23 find the 95 per cent confidence interval for μ and use it to test the hypothesis stated.

13. In Exercise 8.31 calculate the 95 per cent confidence limits for $\mu_x - \mu_y$ where μ_x is the population mean I.Q. of the second school, and μ_y, that of the first school.

14. In Exercise 8.33 construct a 90 per cent confidence interval for $\mu_y - \mu_x$ based on the data given for the two samples. By means of it test the hypothesis $H_0: \mu_y - \mu_x = 0$.

15. In Exercise 8.38 construct the 95 per cent confidence interval for μ_d where $d = x - y$, and x and y designate the readings of instruments C and W respectively.

16. The scores, x, of 63 seniors on a Graduate Record Examination showed $\Sigma x = 34,540$ and $\Sigma x^2 = 19,480,000$. Calculate the unbiased estimate of the population variance of scores. What is your estimate of σ?

17. (a) A sample of 26 has a variance, $s^2 = 2.50$. What is an unbiased estimate of σ^2?

(b) If two samples of sizes 10 and 15 are drawn from the same population and have variances of 2.40 and 2.70 respectively, what is an unbiased estimate of the population variance?

18. Three samples of sizes 10, 20, and 30 are drawn from populations having the same variance, σ^2, and show variances of 12.3, 14.5, and 11.2 respectively. Find an unbiased estimate of σ^2.

19. We know that the variable z, where $z = \dfrac{(\bar{x} - \mu)\sqrt{N}}{\sigma}$ has, in general, a normal distribution. If you did not know σ but did know s, and replaced σ by its approximate point estimate $\sqrt{\hat{\sigma}^2}$, what kind of a variable would z become?

20. Assume that in testing the hypothesis, $\mu_x = \mu_y$, we did not know anything about σ_x^2 and σ_y^2 except that they were equal. Replace the numerators of the denominator of (8.10) by the unbiased estimate of the common variance, σ^2. What familiar formula do you obtain?

21. Prove that when the N's are large, say $N_x + N_y - 2 > 100$, so that $N_x + N_y - 2 = N_x + N_y$ approximately, the estimated variance of the difference of two means given by the denominator of Formula (8.13), may be expressed approximately by

$$\sigma^2_{\bar{x}-\bar{y}} = \frac{s^2_x}{N_y} + \frac{s^2_y}{N_x}.$$

22. The percentage of purity of a medical preparation was tested by means of 5 samples of 10 vials each and produced the following data. Estimate the standard deviation of the population, (a) with the use of the range and (b) without it.

Sample	s^2	R
1	2.72	6.45
2	2.13	4.63
3	2.45	5.31
4	3.06	7.12
5	2.31	5.26

23. From a lot of capsules the contents of five samples of four each were weighed in milligrams. The results follow:

Sample no.	1	2	3	4	5
	62.3	61.9	63.1	62.5	62.5
	62.0	61.8	61.9	62.3	62.1
	62.9	62.0	61.9	62.5	61.8
	62.8	62.2	62.6	61.6	62.9

(a) Calculate the variance of each sample and from these variances estimate the population (lot) variance of the weighed contents of capsules.
(b) Find the range of each sample and from the mean range estimate the lot variance.
(c) How do the estimates made in (a) and (b) compare?

24. If 180 manufactured items were found to have 40 defective, what are the 95 per cent confidence limits for the percentage defective in the population from which the 180 were selected? Use the tables described in Reference 22.

25. In a group of 28 patients 19 reported immediate improvement upon use of a given drug. Find the 95 per cent confidence limits for the percentage of such patients expected to improve. Use the table described in Reference 22.

26. From the ten values of s given in Table 2–3 estimate the population variance of the weights in grains of explosive charges.

THE CHI-SQUARE
DISTRIBUTION

11

"Come, let us see how the squares go."
Anonymous

11.1. An Example

Consider the data of Table 11–1, where the actual frequency distribution that resulted when six pennies were tossed 128 times is compared with the theoretical distribution.

Table 11–1

No. of heads	Probability p_i	Theoretical no. of heads: $128p_i$	Actual no. of heads: f_i
0	1/64	2	2
1	6/64	12	10
2	15/64	30	28
3	20/64	40	44
4	15/64	30	30
5	6/64	12	10
6	1/64	2	4
Sum	1	128	128

The probabilities listed in the second column were obtained from the binomial probability formula

$$\frac{6!}{x!\,(6-x)!}\left(\frac{1}{2}\right)^{x}\left(\frac{1}{2}\right)^{6-x}, \qquad x = 0, 1, 2, 3, 4, 5, 6.$$

By the multinomial formula (Section 9.5) the probability for the complete set of seven frequencies that actually occurred,

$$P = \frac{128!}{2!\,10!\,28!\,44!\,30!\,10!\,4!} \left(\frac{1}{64}\right)^2 \left(\frac{6}{64}\right)^{10} \left(\frac{15}{64}\right)^{28} \left(\frac{20}{64}\right)^{44} \left(\frac{15}{64}\right)^{30} \left(\frac{6}{64}\right)^{10} \left(\frac{1}{64}\right)^4.$$

(11.1)

If we were to ask, "What is the probability for a result as usual as this or less usual?" we would be faced with the task of calculating a sum of probabilities similar to that in (11.1) and this task we would not relish. Just as the (continuous) normal distribution enabled us to compute easily a sum of (discrete) binomial probabilities, so does the (continuous) *Chi-square* (Greek symbol, χ^2) distribution furnish us with a means of approximating a sum of (discrete) multinomial probabilities.

11.2. The Chi-Square Function

The distribution just referred to is very important and is defined by means of the function

$$y = Ce^{-(\chi^2/2)}(\chi^2)^{(n/2)\,-\,1}$$

(11.2)

where C is a constant depending on n, the number of degrees of freedom (Section 8.8). The curve for $n - 6$ is shown in Figure 11–1. It is skew with a range of values from $\chi^2 = 0$ to $\chi^2 = \infty$. When $n = 1$ the curve for χ becomes

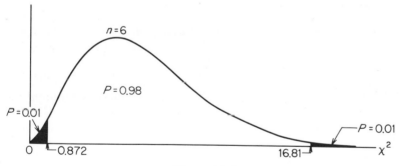

Figure 11–1

the right half of a normal curve with doubled ordinate (Figure 11–2). As n increases, the curve becomes more symmetrical. In fact if we set $z = \sqrt{2\chi^2}$, the resulting distribution of z can be proved to approximate very well, for moderately large values of N, the normal curve with mean at $\sqrt{2n - 1}$. For this reason when $n > 30$ one obtains good practical accuracy by using the normal curve with the standard deviate

$$z = \sqrt{2\chi^2} - \sqrt{2n - 1}.$$

The values of χ^2 for degrees of freedom from $n = 1$ to $n = 30$ have been tabulated for various convenient probability values and may be found in Table F. This table yields the values for the probability, P, that χ^2 exceeds a

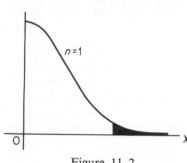

Figure 11–2

given value. The shaded portion of the right tail in Figure 11–1, where $n = 6$, represents the probability, 0.01, that χ^2 exceeds 16.81; the area to the right of the ordinate erected at $\chi^2 = 0.872$ represents the probability, 0.99, that $\chi^2 > 0.872$; hence, the shaded portion of the left tail corresponds to the probability 0.01, that $\chi^2 < 0.872$. We may combine these statements into the double inequality assertion,

$$P(0.872 < \chi^2 < 16.81) = 0.98.$$

A convenient notation for probability statements such as the first two just made is illustrated by the following:

$$\chi^2_{0.01} = 16.81; \qquad \chi^2_{0.99} = 0.872.$$

The former may be read, "Chi-square at the one per cent level equals 16.81."

11.3. Chi-Square and Variance

A well-known theorem of mathematical statistics is the following:

If s^2 is the variance of a sample of N drawn from a normal population with variance, σ^2, then Ns^2/σ^2 has a χ^2-distribution with $N - 1$ degrees of freedom.

Let us consider an application of this important fact.

Example. A certain time factor in a manufacturing operation has a standard deviation of 14 seconds. If a group of 12 workers show a standard deviation of 17 seconds, could we say that this difference in variability is significant on the ten per cent level?

First find the (90 per cent) acceptance region for χ^2 with $n = 11$, $\chi^2_{0.95} = 4.575$ and $\chi^2_{0.05} = 19.68$. Hence

$$P(4.575 < \chi^2 < 19.68) = 0.90.$$

Assuming the distribution of times to be normal we set $\chi^2 = \dfrac{Ns^2}{\sigma^2}$ and substitute $N = 12$, $s = 17$, and $\sigma = 14$ so that

$$\chi^2 = \frac{12 \times 17^2}{14^2} = 17.7.$$

Let H_0 be $\sigma = 14$.

Since this sample value of χ^2 lies within the acceptance interval, we cannot reject H_0.

One can also derive 90 per cent confidence limits for σ^2 from the double inequality

$$4.575 < \frac{12 \times 17^2}{\sigma^2} < 19.68.$$

11.4. Testing a Multinomial Distribution

The data of Table 11–1 will illustrate the title of this section. We reproduce in Table 11–2 the data, slightly altered in form, for a reason to be stated later. [See Section 11.5(b).] In this table the actual frequencies are

Table 11–2

No. of heads	Theoretical frequency: Np_i	Actual frequency: f_i	Difference $f_i - Np_i$
0 or 1	14	12	−2
2	30	28	−2
3	40	44	4
4	30	30	0
5 or 6	14	14	0
Total	128	128	0

to be compared with the theoretical, under the hypothesis that $p = \frac{1}{2}$ for each penny. The distribution is clearly multinomial. Are the experimental or actual outcomes consistent with this hypothesis?

If the experiment of tossing six pennies 128 times were repeated over and over again, each experiment would yield five actual frequencies such as those listed in the third column of Table 11–2. Any frequency could be an integer from 0 to 128 inclusive, provided that the total frequency equalled 128. Thus, there would be an exceedingly large, but not an infinite, number of ways in which the five frequencies could occur, but these frequencies would be subject to the restriction

$$\sum_{i=1}^{5} f_i = 128.$$

At this point we introduce the important fact that the quantity

$$\sum_{i=1}^{m} \frac{(f_i - Np_i)^2}{Np_i}, \tag{11.3}$$

although a discrete variable, has been shown to behave like χ^2 with n degrees of freedom. In (11.3) m is the number of frequencies and in this problem $n = m - 1 = 4$. If we set χ^2 equal to the quantity (11.3), then the number of possible values of χ^2 is finite, its distribution is discrete and representable by means of a histogram. Areas of this histogram can be well approximated by corresponding areas under the true χ^2 curve defined in (11.2). Note that the quantity (11.3) is a sort of pseudo-χ^2, a discrete variable, whereas the χ^2 of (11.2) is, so to speak, the genuine one, the continuous variable. The use of χ^2 in these two senses does not usually cause difficulty.

From the last column of Table 11–2 and from (11.3) we find that

$$\chi^2 = \frac{(-2)^2}{14} + \frac{(-2)^2}{30} + \frac{4^2}{40} + \frac{0^2}{30} + \frac{0^2}{14}$$

$$= 0.82.$$

It is clear that if $\chi^2 = 0$, the agreement between the theoretical and the actual frequencies is perfect. As χ^2 increases, the disparity between the two sets of values increases also. The question that then arises is the following: What is the probability that, in a given sample, the discrepancy between the actual values and the theoretical will lead to a value as large as or larger than 0.82?

The five *cell frequencies* were subject to the single restriction

$$\sum_{i=1}^{5} f_i = 128,$$

hence the number of degrees of freedom, $n = 5 - 1 = 4$. To find the probability, P, that $\chi^2 > 0.82$ for four degrees of freedom we note that

$$\chi^2_{0.95} = 0.711 \quad \text{and} \quad \chi^2_{0.90} = 1.064.$$

Since $0.711 < 0.82 < 1.064$

it follows that $0.90 < P < 0.95.$

Thus, the probability for sample frequencies to deviate from the theoretical at least as much as these do is close to unity. We say, then, that the actual frequencies are in excellent agreement with the theoretical. There is no reason to doubt the validity of the hypothesis that the distribution is a multinomial one with $p = \frac{1}{2}$ for each penny.

11.5. Remarks on the Chi-Square Test

Mathematical considerations involved in the derivation of the χ^2-function and experience with practical problems lie behind the formulation of the following general rules for the applications of χ^2.

(a) Each cell frequency, f, should exceed 5 at least, and should preferably be much larger. When cell frequencies are too small they may be grouped together. This was done in forming Table 11–2 from Table 11–1.

(b) The number, m, of classes or cells should be neither too large nor too small. If $5 \leq m \leq 20$ one is usually on the safe side. Smaller values of m than 5 may be compensated somewhat by ensuring that the f's are larger than 5.

(c) The restrictions imposed on the cell frequencies must be expressible as equations of the first degree in the f's. Thus, in the example

$$f_1 + f_2 + f_3 + f_4 + f_5 = 128.$$

(d) Values of P near unity are sometimes suspect, as the character of the sample or the design of the experiment may be faulty.

11.6. Goodness of Fit to a Normal Curve

The χ^2-test is sometimes applied to graduated data. Consider, for example, the graduated frequency distribution of head lengths shown in Table 7–1 and reproduced here in altered form. In Table 11–3, certain frequencies near the extremities of the table have been combined.

Table 11–3

COMPUTATION OF χ^2 FOR THE FREQUENCY DISTRIBUTION OF HEAD LENGTHS

Class intervals	Actual frequency f_i	Graduated frequency f_i'	$f_i - f_i'$	$(f_i - f_i')^2$	$\dfrac{(f_i - f_i')^2}{f_i'}$
171.5–179.5	12	12.9	−.9	.81	.06
179.5–183.5	29	33.1	−4.1	16.81	.51
183.5–187.5	76	71.3	4.7	22.09	.31
187.5–191.5	104	104.2	−.2	.04	.00
191.5–195.5	110	108.8	1.2	1.44	.01
195.5–199.5	88	77.3	10.7	114.49	1.48
199.5–203.5	30	37.5	−7.5	56.25	1.50
203.5–219.5	13	16.5	−3.5	12.25	.74
Total	462	461.6			4.61

The value of χ^2 is seen to be 4.61. The number of classes is 8, but the number of degrees of freedom is not *one* less, but *three* less than 8, or 5. The reason for this is the fact that the fitted normal curve is determined by means of three statistical measures, N, $\bar{x}$, and s_x^2, where

$$N = \sum_{i=1}^{m} f_i,$$

$$\bar{x} = \frac{1}{N} \sum_{i=1}^{m} f_i x_i,$$

and

$$s_x^2 = \frac{1}{N} \sum_{i=1}^{m} f_i(x_i - \bar{x})^2,$$

all of which satisfy the condition expressed in (d) of the previous section.

Thus three conditions are imposed upon the set of variates composing the frequency distribution. Since, then, $n = 5$, we find from Table F that P is in the neighborhood of 0.47. This means that in about 47 cases out of 100 we should expect a fit as poor as this or poorer. Thus, the hypothesis that head lengths of criminals form a normal distribution seems to be a tenable one, since the probability, 0.47, for as wide a fluctuation is reasonably large.

11.7. 2 × 2 Tables

An example: In a study of "College Affiliation and Political Attitudes," samples of students eligible to vote for the first time in a presidential election were taken from the College of Liberal Arts (CLA) and the College of Business Administration (CBA) of Boston University in 1956.[1] The numbers not in parentheses in Table 11–4 summarize the results of the study.

Table 11–4

	CLA	CBA	Totals
Eisenhower	30 (36.9)	38 (31.1)	68
Stevenson	34 (27.1)	16 (22.9)	50
Totals	64	54	118

Let us test the hypothesis that college affiliation and political attitude are independent of each other. More precisely, the null hypothesis may be stated thus, in words: "The fraction of students in the college population (eligible to vote for the first time) who voted for Eisenhower is essentially the same whether they were students enrolled in the CLA or the CBA."

In the first place this hypothesis may be tested by the method indicated in Section 9.3 and Exercise 14 of Chapter 9. $\hat{p} = 68/118 = 0.577$, $\hat{q} = 0.423$, $p_1 = 30/64 = 0.469$, $p_2 = 38/54 = 0.703$, $N_1 = 64$, and $N_2 = 54$. $H_0:p_x = p_y$; $H_1:p_x \neq p_y$. Formulas (9.8), (9.9), and (9.10) lead us to the value (without the correction factor), $z = 2.57$ and this corresponds to a two-tail probability of about 0.01. The hypothesis is thus rejected for any $\alpha > 0.01$.

The χ^2-distribution may be used to advantage in this problem. There are four categories of students; those from the CLA voting for Eisenhower, those from the CLA voting for Stevenson, those from the CBA voting for Eisenhower, and those from the CBA voting for Stevenson. Estimates of the population proportions in these categories are $^{30}/_{118}$, $^{34}/_{118}$, $^{38}/_{118}$, and $^{16}/_{118}$, or 0.254, 0.288, 0.322, and 0.136 respectively. Suppose that the true proportions were 0.25, 0.30, 0.35, and 0.10. Then the probability for

[1] Levin, M. B. and Nogee, P., "College Affiliation and Political Attitudes," *Boston University Graduate Journal*, Vol. 6, 1958.

outcomes as unusual as those shown in Table 11–4 or more so, could be calculated as the sum of multinomial probabilities which could be well approximated with the aid of the χ^2-function.

We estimate the theoretical frequencies by stating that $^{68}\!/_{118}$ or 57.7 per cent of the entire sample voted for Eisenhower and 42.3 per cent voted for Stevenson. If college affiliation and political attitude are independent of each other then 57.7 per cent of the 64 CLA students, or about 37 of them, on the average, would have voted for Eisenhower. The remaining 27 CLA students would have voted for Stevenson. Likewise 57.7 per cent of the 54 CBA students, or 31 of them would, on the average, vote for Eisenhower and the remaining 23, for Stevenson. These theoretical frequencies, computed to the nearest 0.1, appear in parentheses in the table. Denoting the actual frequencies by f_i and the theoretical, by f_i' and applying Formula (11.3) we have

$$\chi^2 = \sum \frac{(f_i - f_i')^2}{f_i'} \tag{11.4}$$

$$= \frac{(-6.9)^2}{36.9} + \frac{(6.9)^2}{27.1} + \frac{(6.9)^2}{31.1} + \frac{(-6.9)^2}{22.9} = 6.66. \tag{11.5}$$

The number of degrees of freedom, as will be shown in the next paragraph is 1, and from Table F $\chi^2_{0.01} = 6.64$ and $\chi^2_{0.05} = 3.84$. Thus it appears that the hypothesis of independence is just barely refuted for $\alpha = 0.01$ and more definitely refuted for $\alpha = 0.05$.

In this illustration we are imagining that the experiment of sampling a total of 118 students is repeated an infinite number of times and that from these repetitions we select the samples that contain the same marginal frequencies, 64, 54, 68, and 50. Then we inquire, what is the probability that for fixed marginal frequencies, we would obtain the cell frequencies shown in Table 11–4. There are 2×2, or 4 cell frequencies given, subject to the restrictions set by the marginal totals. Each row and column of two cells each must have the appropriate marginal total; also, if we know three marginal totals we know all four, for the complete total is the sum of two adjacent ones. Thus, there are only three restrictions, and $n = 4 - 3 = 1$. Another way of determining n is to note that any single cell frequency may be assigned at will, and then the remaining three are determined by the marginal totals.

11.8. The Yates Correction

When we approximated to a binomial distribution by means of the normal we found it desirable to add or subtract a correction of $\frac{1}{2}$ to the actual frequency being tested. For a similar reason, we make use of a correction due to Yates (1934) for 2×2 tables. The rule may be stated as follows.

Reduce the absolute value of each difference, $|f_i - f_i'|$, by $\frac{1}{2}$ before squaring.

Thus

$$\chi^2 = \sum \frac{[|f_i - f_i'| - \frac{1}{2}]^2}{f_i'}. \tag{11.6}$$

If we apply this correction in the previous example, the numerators in (11.5) become $(6.4)^2$ and $\chi^2 = 5.73$. Thus the hypothesis is *not* refuted at the one per cent level but is refuted for a level, say, of two per cent or higher.

Example. It is suspected that different combinations of temperature and humidity affect the number of defective articles produced in a certain workroom. Do the data below confirm this suspicion?

Table 11–5

		Humidity		
		Low	High	
Temperature	Low	10	4	14
	High	3	17	20
		13	21	34

Let $f_1 = 10$, $f_2 = 4$, $f_3 = 3$, $f_4 = 17$. Then $f_1' = \frac{13}{34} \times 14 = 5.4$, whence, because of the marginal totals, $f_2' = 8.6$, $f_3' = 7.6$, $f_4' = 12.4$. Each difference $f_i - f_i' = 4.6$, hence each corrected difference is $4.6 - 0.5 = 4.1$. Then,

$$\chi^2 = (4.1)^2\left(\frac{1}{5.4} + \frac{1}{8.6} + \frac{1}{7.6} + \frac{1}{12.4}\right) = 8.63.$$

For $n = 1$, $P < 0.01$, hence the null hypothesis that the proportion of defectives in the population when the humidity is low is the same whether the temperature is low or high is untenable. The data confirm our suspicion. It appears that a combination of low temperature with low humidity or high temperature with high humidity produces an excessive number of defectives.

11.9. An Alternative Formula

A 2×2 table may be symbolized as follows:

Table 11–6

	A	$\bar{A}$	Row totals
B	a	b	$a + b = r_1$
$\bar{B}$	c	d	$c + d = r_2$
Column totals	$a + c = c_1$	$b + d = c_2$	$N = a + b + c + d$

Here $\bar{A}$ means "not-A" and $\bar{B}$, "not-B".

It may be proved that

$$\chi^2 \text{ (uncorrected)} = \frac{N(ad - bc)^2}{r_1 r_2 c_1 c_2} ; \tag{11.7}$$

$$\chi^2 \text{ (corrected)} = \frac{N[|ad - bc| - N/2]^2}{r_1 r_2 c_1 c_2} . \tag{11.8}$$

Many statisticians prefer Formulas (11.7) and (11.8) to (11.4) and (11.6).

11.10. Mathematical Models for 2×2 Tables

The discussions of Sections 11.7 and 11.8 were devoted to "approximate tests." However, for 2×2 tables with small cell frequencies, certain "exact tests" are available. The multinomial formula supplied one such test. The subject is quite an extensive one and presents many problems still under investigation. In particular, the mathematical model which forms a basis for the derivation of a test, needs careful elucidation. Mainland's Tables (Reference 22) are particularly useful in dealing with experimental data yielding small cell frequencies. A useful and suggestive contribution to 2×2 tables may be found in *Biometrika* (Reference 29), G. A. Barnard, "Significance Tests for 2×2 Tables," Volume 34, 1947, pp. 123–138.

11.11. $p \times q$ Tables

The method of χ^2 used in fourfold tables may be extended to tables having p rows and q columns. Such a *contingency table* is illustrated in Table 11–7, and the method of analysis follows it. Note that Yates' correction is not in general applicable to other than 2×2 tables.

Table 11–7

CLASSIFICATION OF CANDIDATES FOR ADMISSION TO COLLEGE ON THE
BASIS OF THE COLLEGE'S RATING OF CANDIDATES

	Rating I	Rating II	Rating III	Total
Veterans	19 (22.9)	88 (71.2)	175 (187)	282
Non-veterans	33 (29.1)	76 (91.8)	251 (239)	360
Total	52	164	426	642

The fraction of veterans is $282/642$ or 0.440, from which we estimate the theoretical frequencies shown in parentheses. As soon as two frequencies in any two different columns are known, the remaining ones are determined, hence there are two degrees of freedom. In fact, it is easily proved that the number of degrees of freedom for a $p \times q$ table with marginal totals fixed is $(p - 1)(q - 1)$. Here $p = 2$ and $q = 3$, hence $n = 2$.

$$\chi^2 = \frac{(3.9)^2}{22.9} + \frac{(3.9)^2}{29.1} + \cdots + \frac{(12)^2}{239} = 8.30.$$

Since $\chi^2_{0.02} = 7.82$ and $\chi^2_{0.01} = 9.21$, there are clearly significant differences between veterans and non-veterans on the 2 per cent level. If the percentage of veterans in each of the three ratings is computed, we obtain the values 36.5, 53.6, and 41.0. Apparently there is less variation between Ratings I and III than between any other two.

11.12. The F-Distribution: a Test for $\sigma_1^2 = \sigma_2^2$

Certain tests of the significance of the difference between two means assumed a common population variance, which was estimated from the two sample variances. (Section 8.10.) This assumption is not always valid, so that it may be desirable first to see if the discrepancy between the two sample variances is unusual enough to render this assumption untenable. This is most easily done by means of the variable F, defined as the quotient of two chi-squares:

$$F = \frac{\chi_1^2}{N_1 - 1} \bigg/ \frac{\chi_2^2}{N_2 - 1} , \tag{11.9}$$

where χ_1^2 and χ_2^2 are independently distributed with $N_1 - 1$ and $N_2 - 1$ degrees of freedom respectively. If

$$\chi_1^2 = \frac{N_1 s_1^2}{\sigma_1^2} \quad \text{and} \quad \chi_2^2 = \frac{N_2 s_2^2}{\sigma_2^2} \tag{11.10}$$

are derived from samples drawn independently from two normal populations having equal variances, $\sigma_1^2 = \sigma_2^2$, then (11.9) becomes

$$F = \frac{N_1 s_1^2}{N_1 - 1} \bigg/ \frac{N_2 s_2^2}{N_2 - 1} \tag{11.11}$$

$$= \hat{\sigma}_1^2 / \hat{\sigma}_2^2, \tag{11.12}$$

where $\hat{\sigma}^2$ designates the unbiased estimate of σ^2 and $\hat{\sigma}_1^2$ is always the larger of the two sample variances. (Section 10.4.) Formula (11.12) defines the *variance-ratio* and the symbol F is used in honor of R. A. Fisher.

The distribution of F is well known and is defined by the function

$$y = C F^{\frac{n_1 - 2}{2}} \left(1 + \frac{n_1}{n_2} F \right)^{-\frac{n_1 + n_2}{2}} \tag{11.13}$$

where C is a constant depending upon n_1 and n_2. Here the parameters n_1 and n_2 represent degrees of freedom; $n_1 = N_1 - 1$ and $n_2 = N_2 - 1$, N_1 and N_2 being the sample sizes.

The curve of F is skew with a range from 0 to ∞, and a mean at

$$F = \frac{n_2}{n_2 - 2}.$$

(See Figure 11–3.) It is particularly interesting because of the following facts: (1) If $n_1 = 1$ and n_2 remains finite, the distribution of F is the same as

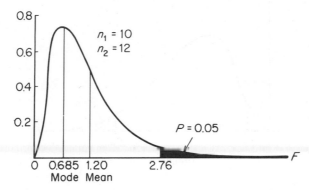

Figure 11–3

that of t^2 (Fisher's t). (2) If n_1 remains finite and n_2 becomes infinite, the F-distribution becomes the Chi-square distribution. (3) If, in the previous case, $n_1 = 1$, the F-distribution reduces to a z^2 distribution where z is distributed normally. Thus, the F-distribution embraces three important distributions.

Tables G of the F-curve give only critical values F_2, corresponding to right-tail areas, that is, to values of $F > 1$. If one should be interested in a left-tail area $(F < 1)$ such that $P(F < F_1) = P(F > F_2) = \alpha$, (Figure 11–4), then F_1 can be found by means of the relation

$$F_1(n_1, n_2) = \frac{1}{F_2(n_2, n_1)}.$$

The proof of this statement is omitted. A test of $H_0: \sigma_1^2 = \sigma_2^2$ versus $H_1: \sigma_1^2 \neq \sigma_2^2$ calls for a two-tail rejection region but we always enter the appropriate table with (n_1, n_2) where n_1 is associated with the larger $\hat{\sigma}^2$ so that $F(n_1, n_2) > 1$. Tables G(a) and G(b) list values of F corresponding to right-tail probabilities of 0.05, 0.025, and 0.01. Any value of $F > 1$ that lies in a right-tail rejection region ensures that the reciprocal $1/F$ lies in the left-tail rejection region, hence we have no need to employ values of $F < 1$.

A large and important field of statistical inquiry is called the *analysis of variance*. Here the alternative to $H_0: \sigma_1^2 = \sigma_2^2$ is always $H_1: \sigma_1^2 > \sigma_2^2$ so that a right-tail rejection region is of sole interest. (See Chapter 14.)

Referring to Table G(a) we find that if $n_1 = 10$ and $n_2 = 12$, the 5 per cent and 1 per cent points for F are 2.76 and 4.30 respectively. Thus $P(F > 2.76) = 0.05$ (Figure 11–3). Another way of stating this fact is by means of the familiar notation $F_{0.05} = 2.76$.

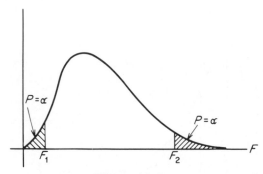

Figure 11–4

Example. In testing for per cent of ash content, 17 tests from one shipment of coal show $s = 2.66$ per cent, and 21 tests from a second shipment show $s = 4.55$ per cent. Test $H_0: \sigma_1^2 = \sigma_2^2$ against $H_1: \sigma_1^2 \neq \sigma_2^2$ with $\alpha = 0.10$.

$$\hat{\sigma}_1^2 = {}^{21}\!/_{20}\ (4.55)^2 = 21.7;$$
$$\hat{\sigma}_2^2 = {}^{17}\!/_{16}\ (2.66)^2 = 7.53;$$
$$F(20, 16) = {}^{21.7}\!/_{7.53} = 2.88.$$

$F_2(20, 16) = 2.28$ for a right-tail probability of 0.05. Since $F = 2.88$ exceeds $F_2 = 2.28$ so that F lies in the right-tail rejection region, we cannot accept H_0.

More complete tables of F, by Merington and Thompson may be found in Volume 33 (1943–1946) of *Biometrika* (Reference 29).

EXERCISES

In the following exercises assume normality if necessary.

1. A sample of 30 drawn from a normal population had a variance of 10. Find the 90 per cent confidence interval for σ^2.

2. The scores on a Graduate Record Examination of 22 seniors from College A showed a standard deviation of 85. Test $H_0: \sigma = 100$ versus $H_1; \sigma < 100$, with $\alpha = 0.04$.

3. The mean purity of a compound is stated to be 94.6 per cent and the standard deviation, 2.13. A sample of 10 from Laboratory Q showed $s = 2.52$. Test H_0; $\sigma = 2.13$ versus H_1; $\sigma \neq 2.13$. Let $\alpha = 0.02$.

4. Assume for a certain age group of American males that systolic blood pressures show a variance of 268. A selected sample of 20 men from this age group had a variance of 313. May one conclude that this group represents a population with $\sigma^2 \neq 268$? Let $\alpha = 0.04$.

5. The variance in the diameter of steel shafts is found to be 0.0052. Would 100 shafts turned out by a certain machinist, with a variance of 0.0065 indicate greater variability than normal for this workman?

6. Refer to Exercise 5.16. A random sample of 100 persons from a New England community showed the following blood groupings: O, 50; A, 35; B, 8; AB, 7. (a) Write, but do not attempt to evaluate, the probability for this particular result. (b) Using $\alpha = 0.10$, decide if this sample is unusual or not.

7. In tossing four pennies 160 times the actual frequencies of occurrence of 0, 1, 2, 3, and 4 heads were 7, 45, 53, 49, and 6 respectively. Discuss this outcome with the aid of the χ^2-test.

Test for goodness of fit any of the frequency distributions in Exercises 8–10 which you have already graduated (according to the wish of your instructor).

8. Thicknesses of washers. Exercise 7.46.

9. Brain-weights of Swedish males. Exercise 7.47.

10. Sizes of shoes worn by college girls. Exercise 7.48.

In each of the Exercises 11–15 test for independence by means of Chi-square.

11. Effect of a new treatment on patients afflicted with a certain disease.

	Recovered	Died	Total
Treated	73	12	85
Not treated	50	21	71
Total	123	33	156

12. Is any relationship between sex and performance indicated in the following data taken from an examination?

	No. right	No. wrong
Boys	56	24
Girls	34	36

13. Influence of the duration of a certain mental disorder on the intelligence quotient.

IQ Duration	Below 70	70 or above	Total
Less than 10 years	62	55	117
More than 10 years	44	26	70
Total	106	81	187

14. Is there a significant difference between Plots 1 and 2?

	No. seeds germinating	No. seeds not germinating
Plot 1	300	100
Plot 2	400	100

15. A drug manufacturing company made a survey of the number of doctors prescribing a certain product. According to the data shown below, does the general practitioner differ materially from the specialist in prescribing this drug?

	Prescribing	Not prescribing
General practitioners	33	204
Specialists	18	47

16. A large ice cream manufacturing company tested its product against that of its competitor by offering to each of 882 individuals two samples of ice cream for one of which a preference was to be stated. The same individual was then given another two samples for one of which a preference was to be stated. Unknown to the taster, the two brands A and B of the first test were the same as those of the second test. The results follow. B represents the brand of the competitor. Discuss these results as fully as possible.

Preference		Frequency
(1)	(2)	
A	A	331
B	B	135
A	B	} 416
B	A	

17. The Pasteur Drug Store found that of the last 160 patrons of its ice cream bar, 70 ordered vanilla ice cream, 42, chocolate, and 48, other flavors. Is this experience consistent with the hypothesis that 50 per cent of its patrons order vanilla, 30 per cent chocolate, and 20 per cent other flavors?

18. In a study of the evaluation of a certain social project, four levels A, B, C, D, of judgment were used. It was expected that, in general, these levels would be given 10, 40, 40, and 10 per cent of the time, respectively. In one study 16 judges showed the following results:

	A	B	C	D
Expected	1.6	6.4	6.4	1.6
Actual	1	7	8	0

It was found that

$$\chi^2 = \frac{(0.6)^2}{1.6} + \frac{(0.6)^2}{6.4} + \frac{(1.6)^2}{6.4} + \frac{(1.6)^2}{1.6} = 2.29.$$

Comment on this method.

19. A professor thinks that students in Monday, Wednesday, Friday classes are more likely to be absent on Mondays and Fridays than on Wednesdays. He noted that in a randomly selected class there were 35 absences on Mondays, 23 on Wednesdays, and 32 on Fridays. Test the hypothesis that absences on the three days in question are about equally numerous.

20. A safety engineer at the $A.B.C.$ Company claimed that accidents were more likely to occur at the beginning and the end of the week and just before lunch and closing time. Do the data shown for a month at the factory of this company bear out his contention?

Day of the week	Hour of day		
	11–12	4–5	Other
Monday	7	10	52
Friday	9	13	57
Other	16	28	112

21. Four closely related species of plants have their seeds mixed in numbers of 10, 30, 40, and 20. A student attempts to separate them, and obtains the numbers shown. Does he do a good job?

Species	Mixture	Results
A	10	8
B	30	37
C	40	20
D	20	35

22. The muzzle velocities (ft per sec) of 7 shells fired from a 75 mm gun showed a variance of 150. The muzzle velocities of 6 shells fired from the same gun, but with a different brand of powder, showed a variance of 120. Is this difference in variability unusual?

23. The pupils of the same age of two different schools were compared for variability of intelligence. A sample of 25 from one school had a variance of 16 in their I.Q.'s, while a sample of an equal number from the other school had a variance of 8. Is there a significant difference in variability?

24. Two lots of steel wire were tested for breaking strength. One lot showed a variance of 230, the other a variance of 492. Discuss the significance of the difference in variances under the assumption that the lots tested numbered
    (a) 7 and 9 respectively;
    (b) 70 and 90 respectively;
    (c) 200 and 200 respectively.

25. Two groups of 10 and 17 patients each in the same stage of a given disease but under different treatments showed mean blood counts of 12,000 and 14,000 respectively, and standard deviations of 1500 and 1000 respectively. Fisher's t-test shows that the difference in blood counts is highly significant. Can this be attributed, in part at least, to a difference in variability?

26. Prove that Formula (11.7) is equivalent to (11.4).

In Exercises 27 and 28 test the hypothesis of independence ($\alpha = 0.05$) with the aid of Mainland's tables (Reference 22) if available.

27.

	B	$\bar{B}$
A	1	9
$\bar{A}$	6	2

28.

	Inoculated	Not inoculated
Attacked	3	7
Not attacked	8	4

29. Apply a χ^2-test of goodness of fit to the data on (a) deaths from horse-kicks in Section 9.4; (b) flying bombs, Exercise 9.26. Note that the restrictions consist in the given values, $\Sigma f = N$ and $x = \dfrac{1}{N} \Sigma fx$. Since, theoretically, $\mu = \sigma^2$ and we use x as an estimate for μ, there are two restrictions. The last two classes in each distribution should be combined.

CURVE FITTING: REGRESSION

12

"An equation is the most serious and important thing in
mathematics."
Sir Oliver Lodge
Easy Mathematics, 1906

12.1. Introduction

The accompanying graphs exhibit markedly different configurations of
plotted points. In Figure 12–1, the points appear to lie fairly well along a
straight line; in Figure 12–2, they appear to lie on or near a curve known as a
parabola; and in Figure 12–3, they do not appear to lie along any recognizable
curve. Configurations of plotted points may or may not indicate trends or
relationships in the data from which they arise. It is often the duty of the
statistician to discover and to measure such trends or relationships when
they exist. This means that he must solve a problem in curve fitting. Section
7.11 dealt with such a problem, but there the type of curve was defined in
advance by means of Equation (7.3). From the given data, $\bar{x}$, s, and the area,
Nk, were computed so that the constants associated with the equation could
be estimated. There remained only the problem of actually constructing the
curve.

In this chapter we are concerned with the more general problem of fitting
an equation or a curve to data involving a finite number of *paired values*. A
frequency distribution pairing variates, x_i, with corresponding frequencies,
f_i, is but one example of such data. Statistical items may be arranged in
other forms. A *time series* pairs one set of variates with intervals or
instants of time; Figure 12–3 pairs scores in Mathematics with scores in
English.

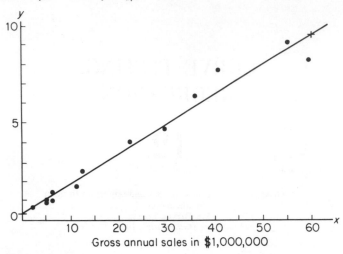

Fig. 12–1.   End-of-year inventory and gross annual sales of 13 U.S. drug store chains for 1954. (*American Druggist*, May 9, 1955, p. 12.)

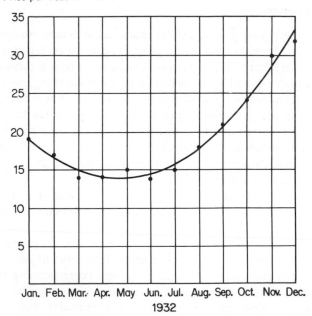

Fig. 12–2.   Average wholesale price per dozen of eggs, Boston, 1932.

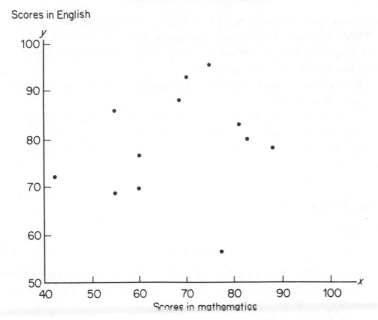

Fig. 12–3. Scores of a group of freshmen in mathematics and English.

Statistical problems centered about curve fitting may be resolved into three parts.

(1) *The statistician must decide what kind of curve to fit to the data.* The decision is often a matter of common sense and good judgment. The best-fitting curve is sometimes too complicated for practical use and must be replaced by a simpler one. Perhaps one portion of the data requires one curve while the remainder demands another. It will suffice now to say that the plotted points themselves most frequently give the clue to the type of curve required.

(2) *The statistician must calculate the constants involved in the equation of the curve selected to fit the data.* This equation is usually written in the form $y = f(x)$, where $f(x)$ is some function of x.

For example, the first-degree equation:

$$y = a + bx \qquad (12.1)$$

defines a straight line. If we selected this equation as the one best fitting the data of Figure 12–1, it would be necessary to calculate the values of the constants a and b which characterize the particular straight line we seek. Constants, such as these, which characterize an equation of a given type are called *parameters*. The parameters of the normal frequency curve (7.3) are μ and σ.

The second part of the general problem is, therefore, to calculate parameters. When this has been done, the curve may be constructed. Good judgment requires that we select a curve whose parameters are not too numerous nor too difficult to compute.

(3) *The statistician may interpret the results by means of explanations, estimates, and predictions.*

There is an infinite variety of curves in mathematics; yet for the purposes of statistical curve fitting, the curves used are relatively limited in type. Fortunately for us, the straight line is the simplest and one of the most important curves used.

12.2. The Straight Line

The equation:

$$y = a + bx \qquad (12.2)$$

is an equation of the first degree in x and y. It can be proved to define a straight line. For this reason it is called a *linear equation*. The point (x_i, y_i)

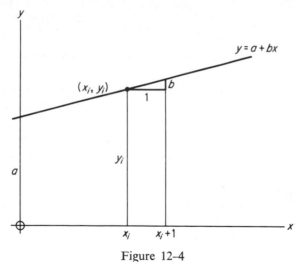

Figure 12–4

lies on the line (12.2) if and only if its coordinates satisfy the equation of the line, that is, if and only if

$$y_i = a + bx_i,$$

or $$y_i - (a + bx_i) = 0.$$

When $x = 0$, $y = a$, hence the line crosses the y-axis at a units from the origin, O (Figure 12–4). a is called the *y-intercept* of the line (12.2).

From Figure 12–4 we can find a simple but significant meaning for the parameter b. For every unit increase in x, y changes by b units; thus b measures the steepness of the line and, for this reason, is called its *slope*. b is also termed the *regression coefficient*. When b is positive the line ascends from left to right, when b is negative the line descends from left to right.

12.3. Fitting a Straight Line by the Method of Least Squares

The selection of a straight line rather than some other curve to fit a set of points is usually made upon the basis of the appearance of the plotted points themselves. One would not hesitate to select a straight line for those of Figure 12–1. There are, however, occasions when we desire to find best-fitting straight lines for points such as those of Figure 12–3.

"Fitting a straight line" usually means finding the values of the parameters a and b of the straight line (12.2) as well as actually constructing the line itself. The "least squares" method assumes that the best-fitting line is the one for which the sum of the squares of the vertical distances of the points (x_i, y_i) from the line is a minimum. It is in very common use and possesses the advantage that it is applicable to more general cases.

If

$$y = a + bx \tag{12.3}$$

is the equation of this line, the ordinate of any point, Q_i, on the line vertically above or below a given point, P_i, can be found by substituting the abscissa, x_i, in the right-hand member of (12.3). The two coordinates of Q_i will be $(x_i, a + bx_i)$ (Figure 12–5). The vertical distance, e_i, from the line of any point P_i with coordinates (x_i, y_i) will therefore be given by the equation:

$$e_i = y_i - (a + bx_i). \tag{12.4}$$

We may say that e_i represents the difference between the *actual* ordinate, y_i, of a point and its *theoretical* ordinate, $a + bx_i$. The quantity e_i is often called a *residual* or *error*. It may be positive or negative.

The best-fitting line is that line for which the sum of the squares, Σe_i^2, is a minimum. Our problem is to find the values of a and b which make Σe_i^2 a minimum. It will be shown in the next section that these values are given by the formulas,

$$b = \frac{N \Sigma x_i y_i - \Sigma x_i \, \Sigma y_i}{N \Sigma x_i^2 - (\Sigma x_i)^2}, \tag{12.5}$$

$$a = \frac{\Sigma x_i^2 \, \Sigma y_i - \Sigma x_i \, \Sigma x_i y_i}{N \Sigma x_i^2 - (\Sigma x_i)^2} \tag{12.6}$$

$$= \bar{y} - b\bar{x}, \tag{12.7}$$

where N is the number of points (x_i, y_i).

It will be useful here to prove that (12.6) is equivalent to (12.7). To do this let us substitute $\bar{x}$ for x in Equation (12.3) and see what the corresponding value of y is. First let us note that the common denominator of (12.5) and (12.6) is

$$N\Sigma x^2 - (\Sigma x)^2 = N^2 s_x^2.$$ **(12.8)**

(See Section 2.10.)

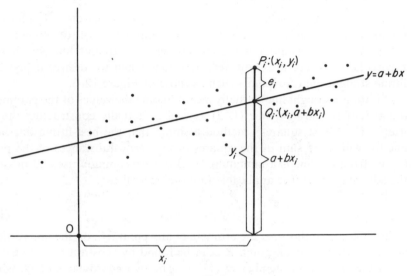

Figure 12–5

If

$$y = a + b\bar{x}$$

then by (12.5) and (12.6)

$$y = \frac{\Sigma x^2 \Sigma y - \Sigma x \Sigma xy}{N^2 s_x^2} + \frac{N\Sigma xy - \Sigma x \Sigma y}{N^2 s_x^2}\bar{x}.$$

But since $\bar{x} = \dfrac{1}{N}\Sigma x$,

$$y = \frac{\Sigma x^2 \Sigma y - \Sigma x \Sigma xy + \Sigma xy \Sigma x - \dfrac{1}{N}(\Sigma x)^2 \Sigma y}{N^2 s_x^2}$$

$$= \frac{\Sigma y\left[\Sigma x^2 - \dfrac{1}{N}(\Sigma x)^2\right]}{N^2 s_x^2}$$

$$= \frac{\Sigma y \cdot N s_x^2}{N^2 s_x^2} = \bar{y}.$$

Thus the means $(\bar{x}, \bar{y})$ determine a point that lies on the fitted line, $y = a + bx$, so that

$$\bar{y} = a + b\bar{x} \qquad (12.9)$$

always, and this is equivalent to Formula (12.7).

12.4. An Example

The tabulation in Table 12–1 shows the data from which Figure 12–1 was constructed.

Table 12–1

Drug store chain	Sales in $1,000,000 x	Inventory in $1,000,000 y	x^2	y^2	xy
Crown	12.2	1.8	148.84	3.24	21.96
Cunningham	41.0	7.8	1681.00	60.84	319.80
Dow	5.4	0.9	29.16	0.81	4.86
Gallagher	13.0	2.6	169.00	6.76	33.80
Gray	22.6	4.1	510.76	16.81	92.64
Katz	35.9	6.4	1288.81	40.96	229.76
Kinsel	7.2	1.3	51.84	1.69	9.36
Parkview	5.2	0.9	27.04	0.81	4.68
Peoples	55.0	9.1	3025.00	82.81	500.50
Reed	2.4	0.7	5.76	0.49	1.68
Sommers	6.8	1.5	46.24	2.25	10.20
Sun Ray	29.6	4.7	876.16	22.09	139.12
Whelan	58.7	8.2	3445.69	67.24	481.34
Totals	295.0	50.0	11,305.30	306.80	1849.70

Formula (12.5) requires the calculation of Σx, Σy, Σx^2, and Σxy. These are found as column totals in Table 12–1. From (12.5) we find that

$$b = \frac{(13 \times 1849.70) - (295.0 \times 50.0)}{(13 \times 11,305.30) - (295.0)^2} = \frac{9296}{59,944} = 0.155,$$

and from (12.7), that

$$a = \frac{50.0}{13} - 0.155 \times \frac{295.0}{13} = 0.33,$$

whence the equation of the least squares line becomes

$$y = 0.33 + 0.155x. \qquad (12.10)$$

Thus we have fitted, *algebraically*, the line (12.10) to the data. To construct the line, that is, to fit it *geometrically* to the points of Figure 12–1, we locate two convenient points on the line. For example, if $x = 0$, $y = 0.33$, and if $x = 60$, $y = 0.33 + (0.155 \times 60) = 9.63$. The two points (0, 0.33) and (60, 9.63) are marked with crosses on Figure 12–1 and determine the line shown there.

If we may consider this fitted line to be representative of the population of drug store chains then we may draw inferences like the following:

(1) The slope of the line is 0.155, hence for every \$1,000,000 increase in annual gross sales, there is, on an average, an increase of approximately \$155,000 in end-of-the-year inventory.

(2) If a chain has annual sales of \$60,000,000 its end-of-year inventory is estimated to be close to \$9,630,000, as seen above.

A test for the validity of such conclusions will be described in Section 12.10.

As another illustration of the use of a fitted line consider the equation

$$y = 58.0 - 1.50x, \tag{12.11}$$

where y represents the yearly infant mortality rate (number of deaths per 1000) in New York City and $x = 0, 1, 2, \ldots$, correspond to the years 1930, 1931, 1932, ..., respectively. We may make statements like the following:

(1) There was an average yearly *decrease* of 1.50 deaths per 1000 among infants.

(2) In 1935 $(x = 5)$ the rate is estimated to have been $y = 58 - (1.50 \times 5)$, or 50.5 per 1000.

(3) To find, for example, the year in which the rate was 43.0, we set $43.0 = 58.0 - 1.50x$, whence $x = 10$ and this corresponds to the year 1940.

All of the above conclusions assume that the trend indicated by the line (12.11) continues through the range of prediction.

12.5. The Derivation of the Least Squares Formulas

Students familiar with calculus should find the following derivation of Formulas (12.5) and (12.6) not difficult. Let

$$S = \Sigma e_i^2 = \Sigma[(y_i - (a + bx_i)]^2.$$

[See Formula (12.4).] To find the values of the two parameters a and b that minimize this sum, we set the partial derivatives with respect to a and b, equal to zero, that is, we differentiate S, first, with respect to a only and then with respect to b only. Note that x_i and y_i are constants, the data given.

$$\left. \begin{array}{l} \dfrac{\partial S}{\partial a} = 2\Sigma(y_i - a - bx_i)(-1) = 0, \\[2mm] \dfrac{\partial S}{\partial b} = 2\Sigma(y_i - a - bx_i)(-x_i) = 0. \end{array} \right\}$$

These two equations reduce to

$$\left. \begin{array}{l} \Sigma y_i - Na - b\Sigma x_i = 0, \\[2mm] \Sigma x_i y_i - a\Sigma x_i - b\Sigma x_i^2 = 0. \end{array} \right\}$$

The values that satisfy this pair of linear equations in a and b is found by elementary algebra to be given by (12.5) and (12.6).

12.6. Time Series

Frequently the x_i's represent end- or mid-points of equal intervals of time. Thus x_1, x_2, ..., x_{12}, might represent the final days of January, February, ..., December, respectively, and y_1, y_2, ..., y_{12}, the total cash sales of a store for the corresponding months; or the x_i's might represent the hours of the day, $x_1 = 9$ (A.M.), $x_2 = 10$ (A.M.), ..., $x_9 = 5$ (P.M.) and the y_i's, the hourly readings of a pressure gauge. In such cases, or in any cases where the x intervals are constant, it may be desirable to simplify the computation by coding. One could let

$$t_i = \frac{x_i - \bar{x}}{k},$$

where k is the length of the interval. Then Formulas (12.5) and (12.7) become

$$b' = \frac{N \Sigma t_i y_i - \Sigma t_i \Sigma y_i}{N \Sigma t_i^2 - (\Sigma t_i)^2}, \qquad (12.12)$$

$$a' = \bar{y} - b'\bar{t}. \qquad (12.13)$$

Since $\Sigma(x_i - \bar{x}) = 0$ (Section 2.4), $\Sigma t_i = 0$ and $\bar{t} = 0$. Then (12.12) and (12.13) reduce to

$$b' = \frac{\Sigma t_i y_i}{\Sigma t_i^2}, \qquad (12.14)$$

$$a' = \bar{y}. \qquad (12.15)$$

When the b' and a' are found, the least squares line in y and t,

$$y = a' + b't,$$

is converted into the desired line in x and y;

$$y = a' + b'\left(\frac{x_i - \bar{x}}{k}\right),$$

which can be written in the form

$$y = a + bx,$$

where $a = a' - \dfrac{b'\bar{x}}{k}$, and $b = \dfrac{b'}{k}$.

12.7. Polynomial Curve Fitting

A *rational integral function of x* or *polynomial in x* is a function of the form

$$a_0 + a_1x + a_2x^2 + \cdots + a_{n-1}x^{n-1} + a_nx^n,$$

where the a's are constant coefficients and n is a positive integer. The corresponding equation is:

$$y = a_0 + a_1x + a_2x^2 + \cdots + a_{n-1}x^{n-1} + a_nx^n.$$

If $n = 1$, the polynomial is linear; if $n = 2$, quadratic; and so on.

The method of least squares may be used to fit a polynomial to data. If (x_i, y_i), where $i = 1, 2, \ldots, N$, are the given pairs of values, then we have the problem of finding the values of $a_0, a_1, \ldots, a_n$, that minimize the sum

$$S = \Sigma e_i^2 = \Sigma[y_i - (a_0 + a_1x_i + \cdots + a_nx_i^n)]^2.$$

There result $n + 1$ equations. The labor of calculation is considerably increased for values of $n > 2$. A widely used systematic method of solving for the a's and checking the calculations is that named after its originator, M. H. Doolittle. When there is doubt concerning the degree of the polynomial to be used, an efficient technique for "trying out" a set of polynomials of increasing degree exists in the construction of *orthogonal polynomials*.

12.8. Regression

The word regression came into statistical use when Sir Francis Galton (1822–1911), in investigations into heredity, found that "the sons of fathers who deviate x inches from the mean height of all fathers, themselves deviate from the mean height of all sons by less than x inches." Galton termed this a "regression to mediocrity." Today the word has forsaken this narrow meaning because statisticians have used it in a broader sense to connote many functional relationships. In studying the simplest kind of relation between two statistical variables, x and y, we make use of the dot diagram and a least squares line. Such a line can be termed a *regression line*. In less simple cases we make use of *regression curves*, as, for example, polynomials, or we may study the interrelations among three or more variables by means of regression equations involving these variables.

In many curve fitting or regression problems we know or suspect that one variable x, is the cause, in part, at least, of the variation in the other variable y. For example, we note that an increase in gross annual sales increases the end-of-year inventory (Figure 12–1); a change in the concentration of an insecticide produces a change in the per cent of insects killed; an increase in the time spent on remedial reading results in an increase in reading speed, and so on. The line of regression of y on x, the causal variable, affords a method of making estimates or predictions of y for a given x. (See also Section 12.11.)

12.9. Explained and Unexplained Variance

The scatter diagram of Figure 12–6 exhibits no upward or downward trend or regression. Obviously, the value of y is not influenced by the value of x. The overall variance of the y's, designated as usual by s_y^2, is the mean of the squares of the lengths of the vertical line segments shown in the figure.

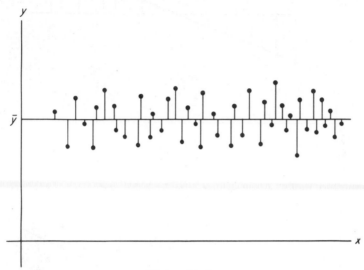

Figure 12–6

In Figure 12–7 *all* the points lie on the line of regression, hence the variability of the y's can be completely explained as due to the regression. In fact, each y is defined by means of the linear equation

$$y = a + bx.$$

The variance s_y^2 is again the mean of the squares of the lengths of the vertical line segments shown in Figure 12–7. In Figure 12–6 none of the variance can be thus explained.

The configuration of points shown in Figure 12–8 indicates that some but not all of the variability among the y's can be explained by regression. If the points all lay on the line as in Figure 12–7 we could account for all of the variance. We measure the failure of the points to lie on the line by means of the distances, e_i, of the points from the line of regression. These afford a measure of the variability that regression does not explain.

We distinguish between two values of y corresponding to x_i, (1) the observed y_i, given by the data and (2) the estimated y_i' calculated from the equation of the regression line, so that

$$y_i' = a + bx_i. \tag{12.16}$$

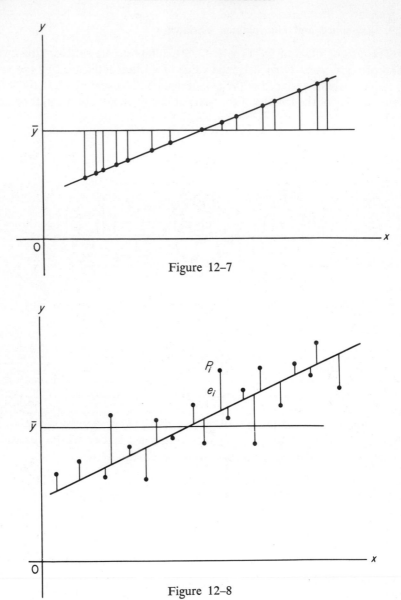

Figure 12–7

Figure 12–8

Thus (x_i, y_i') defines a point lying on the line. By Equation (12.4)

$$e_i = y_i - (a + bx_i)$$

$$= y_i - y_i'. \qquad \textbf{(12.17)}$$

The mean of the estimated y's

$$\bar{y}' = \frac{1}{N} \Sigma y_i'$$

$$= \frac{1}{N} \Sigma (a + bx_i)$$

$$= \frac{1}{N} \Sigma a + b \cdot \frac{1}{N} \Sigma x_i.$$

$$= a + b\bar{x},$$

hence by (12.9) $\qquad\qquad \bar{y}' = \bar{y}.$ $\qquad\qquad$ (12.18)

Thus the mean of the estimated y's equals the mean of the observed y's.

Next $\qquad\qquad \Sigma e_i = \Sigma (y_i - y_i')$

$$= \Sigma y_i - \Sigma y_i'$$

$$= N\bar{y} - N\bar{y}'$$

so that by (12.18) $\qquad\qquad \Sigma e_i = 0.$ $\qquad\qquad$ (12.19)

Thus $\qquad\qquad \bar{e} = \frac{1}{N} \Sigma e_i = 0.$

It follows that the sum of the vertical distances of the points (x_i, y_i) *above* the line (taken as positive) must equal the corresponding sum for the points *below* the line (taken as negative). This fact is useful in fitting a straight line to data visually, rather than by the method of least squares. Frequently a set of points lies close to a straight line as in Figure 12–1, and a line fitted visually may be accurate enough for many purposes. A taut string, a transparent ruler, or a thin ruler placed on edge is located so that the sums mentioned above appear to be equal, and then a line is drawn. The co-ordinates of (x_1, y_1) and (x_2, y_2) of two convenient, widely spaced points on the line are estimated and substituted in the equation $y = a + bx$. This operation produces two equations in the two unknowns, a and b, which can be readily solved for these parameters.

The mean of the squares of the vertical distances e_i, is called the *variance of the errors of estimate* and is designated by s_e^2. Thus,

$$s_e^2 = \frac{1}{N} \Sigma e_i^2.$$ $\qquad\qquad$ (12.20)

The quantity, s_e, is called the *standard error of estimate*.

The variance of the estimated y's

$$s_{y'}^2 = \frac{1}{N} \Sigma (y_i' - \bar{y}')^2$$

$$= \frac{1}{N} \Sigma (y_i' - \bar{y})^2.$$ $\qquad\qquad$ (12.21)

$s_{y'}$ is called the *standard error of the estimated values*.

The following interesting and important relation will now be proved.

$$s_y^2 = s_e^2 + s_{y'}^2. \tag{12.22}$$

$$s_e^2 = \frac{1}{N}\Sigma[y_i - (a + bx_i)]^2.$$

From (12.7)

$$s_e^2 = \frac{1}{N}\Sigma[(y_i - \bar{y}) - (bx_i - b\bar{x})]^2$$

$$= \frac{1}{N}\Sigma[(y_i - \bar{y})^2 - 2b(y_i - \bar{y})(x_i - \bar{x}) + b^2(x_i - \bar{x})^2]$$

$$= \frac{1}{N}\Sigma(y_i - \bar{y})^2 - \frac{2b}{N}\Sigma(x_i - \bar{x})(y_i - \bar{y}) + \frac{b^2}{N}\Sigma(x_i - \bar{x})^2$$

$$= s_y^2 - 2b \cdot \frac{1}{N}\Sigma(x_i - \bar{x})(y_i - \bar{y}) + b^2 s_x^2. \tag{12.23}$$

Let us examine the middle term of the expression on the right.

$$\frac{1}{N}\Sigma(x_i - \bar{x})(y_i - \bar{y}) = \frac{1}{N}\Sigma(x_iy_i - x_i\bar{y} - \bar{x}y_i + \bar{x}\bar{y})$$

$$= \frac{1}{N}\Sigma x_iy_i - \bar{y}\cdot\frac{1}{N}\Sigma x_i - \bar{x}\cdot\frac{1}{N}\Sigma y_i + \frac{1}{N}\Sigma\bar{x}\bar{y}$$

$$= \frac{1}{N}\Sigma x_iy_i - \bar{y}\bar{x} - \bar{x}\bar{y} + \bar{x}\bar{y}$$

$$= \frac{1}{N}\Sigma x_iy_i - \bar{x}\bar{y}$$

$$= \frac{1}{N}\Sigma x_iy_i - \frac{\Sigma x_i}{N}\frac{\Sigma y_i}{N}$$

$$= \frac{N\Sigma x_iy_i - \Sigma x_i\Sigma y_i}{N^2}. \tag{12.24}$$

But from (12.5) and (12.8)

$$b = \frac{N\Sigma x_iy_i - \Sigma x_i\Sigma y_i}{N^2 s_x^2} \tag{12.25}$$

therefore

$$bs_x^2 = \frac{N\Sigma x_iy_i - \Sigma x_i\Sigma y_i}{N^2}$$

hence by (12.24)

$$bs_x^2 = \frac{1}{N}\Sigma(x_i - \bar{x})(y_i - \bar{y}). \tag{12.26}$$

Thus (12.23) becomes

$$s_e^2 = s_y^2 - 2b^2 s_x^2 + b^2 s_x^2$$

or

$$s_e^2 = s_y^2 - b^2 s_x^2. \tag{12.27}$$

Next

$$s_{y'}^2 = \frac{1}{N}\Sigma(y_i' - \bar{y})^2.$$

By (12.16) and (12.9)

$$s_{y'}^2 = \frac{1}{N}\Sigma[(a + bx_i) - (a + b\bar{x})]^2$$

$$= \frac{1}{N}\Sigma b^2(x_i - \bar{x})^2$$

$$s_{y'}^2 = b^2 s_x^2 . \tag{12.28}$$

If, then we add (12.27) to (12.28) we obtain (12.22).

It is now evident that the overall variance, s_y^2, may be decomposed into two parts, the *explained variance*, $s_{y'}^2$, and the *unexplained variance*, s_e^2. The first part is also called the *variance of the estimated values*; the second, the *error* or *residual variance*. Formulas (12.27) and (12.28) are useful in computing these component variances.

12.10. A Test for Linearity

A statistician examining the dot diagram of Figure 12–1 would feel confident that a straight line is appropriate for measuring regression. His opinion might be reinforced by business experience, by a knowledge that inventory is, to a large degree, influenced by sales. If there is doubt, however, concerning the adequacy of a straight line, a simple test exists for the hypothesis that the regression is sufficiently linear to enable valid estimates to be made. The F-distribution (Section 11.12) provides such a test. It can be proved that

$$F = \frac{s_{y'}^2/1}{s_e^2/N - 2}, \tag{12.29}$$

where there is one degree of freedom associated with the explained variance, $s_{y'}^2$, and $N - 2$ degrees of freedom, with the unexplained variance, s_e^2. Essentially we are testing to see if the explained variance is significantly greater than the unexplained. If the latter were unduly large, estimates of y based on the line would not be trustworthy.

EXAMPLE 1.   From the data of Table 12–1 we find $\bar{x} = 22.7$, $\bar{y} = 3.85$, $s_x^2 = 354$, and $s_y^2 = 8.85$. Since $b = 0.155$

$$s_{y'}^2 = (0.155)^2 \times 354 = 8.50, \text{ by (12.28)};$$

$$s_e^2 = 8.85 - 8.50 = 0.35, \text{ by (12.27)}.$$

Then $$F = \frac{8.50}{0.35/11} = 267.$$

By Table G(a), $F_{0.01}(1, 11) = 9.65$. The computed F is, therefore, extremely significant; the variation in y is almost entirely explained by regression.

EXAMPLE 2. For the data of Exercise 12.16 one finds that $N = 12$, $b = 0.123$, $s_{y'}^2 = 2.74$, and $s_e^2 = 104.3$. Then $F = 0.262$. Since $F < 1$, the assumption of linear regression is not valid.

  The test of this section is equivalent to that of the next. The student may show easily that the F of formula (12.29) equals the t^2 of formula (12.30) when $\beta = 0$. The t-test of the following section is generally more useful.

12.11. Testing Hypotheses Concerning Slopes

  It is often desirable to determine if the difference between the sample regression coefficient, b, and an assumed population regression coefficient, β, is such as to be interpreted as ordinary sampling variation or as a denial of such variation. Frequently we may prefer to find a confidence interval for β.

  If we assume that the $x_1, x_2, ..., x_N$, of the sample are constants, not varying from experiment to experiment, but that the corresponding $y_1, y_2, ...,$ y_N, are random variables normally distributed about the population regression line

$$y = \alpha + \beta x$$

with a common variance, σ_e^2, then the variable t defined by the formula

$$t = (b - \beta)\sqrt{\frac{(N - 2)s_x^2}{s_e^2}} \tag{12.30}$$

will have a Student t-distribution with $N - 2$ degrees of freedom. The assumption implies that the values of x are selected in advance by the experimenter as those best suited or most convenient for his investigation and that replications of the experiment would always involve the same set of values of x. The most common form of test arises when we assume $\beta = 0$.

 Example. For the sales and inventory data of Section 12.4, let us find the 90 per cent confidence interval for β.

  One can verify that $N = 13$, $b = 0.155$, $s_x^2 = 354$, and $s_e^2 = 0.35$. Then for 11 degrees of freedom, $t_{0.05} = 1.80$, so

$$\pm 1.80 = (0.155 - \beta)\sqrt{\frac{11 \times 354}{0.35}}$$

whence $0.14 < \beta < 0.17.$

Obviously this is inconsistent with the hypothesis that $\beta = 0$. Thus there is some positive regression.

Given two regression coefficients, b_1 and b_2, and assuming that the population residual variances $\sigma_{e_1}^2$ and $\sigma_{e_1}^2$, are the same, we may test the hypothesis $\beta_1 = \beta_2$, by means of the statistic

$$t = \frac{b_1 - b_2}{\left[\dfrac{N_1 s_{e_1}^2 + N_2 s_{e_2}^2}{N_1 + N_2 - 4}\left(\dfrac{1}{N_1 s_{x_1}^2} + \dfrac{1}{N_2 s_{x_2}^2}\right)\right]^{\frac{1}{2}}} \qquad (12.31)$$

which has a Student t-distribution with $N_1 + N_2 - 4$ degrees of freedom.

EXERCISES

1. Given the following set of points: (1, 2), (4, 5), (7, 7), (10, 12), (13, 13).
   (a) Construct a scatter diagram. (b) Find the equation of the least squares line fitting these points. (c) Construct the line.

2. A group of 5 students took tests before and after training and obtained the following scores:

   Before, x: 8 10 10 15 20
   After, y: 10 12 15 20 20

   Find by the method of least squares the equation of the line fitting these data. Plot the points and the line.

3. When the temperatures, y, in Fahrenheit degrees were plotted against the number, x, of chirps per minute of crickets, the plotted points were found to lie very close to a straight line. The equation of the best fitting line was found to be $y = 40 + 0.25x$. (a) What would you estimate the temperature to be when a cricket is making 120 chirps per minute? (b) How much faster would a cricket chirp if the temperature increased $1°$ F?

4. (a) Fit a least squares line to the points (3, 5), (5, 7), (1, 1), (6, 9), (−2, −3), (3, 4), (4, 6), (−1, −1), (1, 2), (2, 3).
   (b) Calculate the explained and unexplained variances of the y's. What percentage of the total variance is explained by regression?

5. Systematic efforts to reduce the percentage of defective articles produced by one manufacturing unit of a certain plant showed the following results for the first 12 weeks.

Week	1	2	3	4	5	6
Per cent defective	7	6.5	7	6	6	5.5

Week	7	8	9	10	11	12
Per cent defective	6	5	5.5	4	5	4.5

Fit a least squares line to the data and from its equation estimate the average weekly reduction in the percentage of defectives.

6. The temperature readings (y) at regular time intervals (x) were as follows:

Time (minutes)	0	1	2	3	4	5	6
Temperature	70	77	92	118	136	143	155

Find the equation of the regression line and plot this line on a dot diagram of the data. Interpret b.

7. The scores, x, of 22 students in a calculus class on the first hour-quiz and on their final examination, y, yielded $\Sigma x = 1717$, $\Sigma y = 1408$, $\Sigma xy = 112{,}854$, $\Sigma x^2 = 138{,}341$, $\Sigma y^2 = 98{,}818$. Find the equation of the least squares line fitting these data.

8. The mental ages, x, and the scores on a test, y, of a group of 12 boys were as follows:

$$(5, 0), \ (5, 5), \ (7, 8), \ (8, 10), \ (8, 18), \ (9, 20), \ (10, 30), \ (11, 40), \ (12, 35),$$
$$(13, 43), \ (14, 50), \ (15, 50).$$

(a) Find the slope of the least squares line. What information about ages and scores does the value of the slope give you? (b) What percentage of the variance of the test scores can be explained by the regression?

9. Given the number of mice dying per group of 10 from various doses of a drug:

Dosage (x):	50	56	62	70	80
No. dying (y):	0	4	5	6	9

From the equation of the least squares line, estimate the *median lethal dose* (MLD), that is, the dose that will kill just half of the mice.

10. Various doses of a drug are administered to groups of 10 mice and the number dying per dose recorded. The equation of the least squares line was found to be $y = -12.1 + 0.265x$, where x is the dose and y, the number dying. A fixed unknown dose of the same drug is administered to 5 groups of 10 mice each and showed the following numbers dead: 4, 5, 7, 5, 6. Estimate the strength of the dose.

11. In a laboratory the times in minutes, x, required by a group of girls to perform a certain operation were recorded. The girls were then retrained to perform the same operation differently and the times, y, required were noted. Points were plotted to correspond to the two times for each girl and a straight line fitted to them. Its equation is $y = -2.3 + 0.84x$. State two important conclusions that one may draw from this equation.

In Exercises 12–14 fit visually a straight line to each of the sets of points given.

12. $(3, 5), \ (5, 7), \ (1, 1), \ (6, 9), \ (-2, -3), \ (3, 4), \ (4, 6), \ (-1, -1), \ (1, 2), \ (2, 3).$

13. The mental ages, x, and the scores, y, of 12 boys on a test.

x: 5 5 7 8 8 9 10 11 12 13 14 15

y: 0 5 8 10 18 20 30 40 35 43 50 50

14. Number of shares outstanding in the Massachusetts Investors Trust. From your graph estimate the number of shares outstanding in 1960.

Year	1925	1930	1935	1940
No. shares	458,607	3,050,859	19,520,244	35,081,142

Year	1945	1950	1955
No. shares	43,495,344	65,333,604	87,458,325

15. Test for linear regression the data in (a) Ex. 1; (b) Ex. 4; (c) Ex. 8.

16. A group of freshmen received the following scores in Mathematics, (x), and English, (y), examinations (see Fig. 12–3.):

x: 43 55 55 60 60 68 71 76 78 83 82 88

y: 71 84 68 76 69 87 92 93 55 77 82 77.

Verify the values of b, $s_{y'}^2$, and s_e^2 given in Example 2 of Section 12.10.

17. In Exercise 8 find the 95 per cent confidence interval for β. By means of it test the hypothesis that $\beta = 3$.

18. In Exercise 7 test the hypothesis that $\beta = 0$.

19. The following data were obtained on the hardness, x, and the tensile strength, y, of five specimens of castings:

x: 53 70 55 53 69

y: 29 34 30 31 36

Find the 90 per cent confidence limits for the regression coefficient, β, of the population regression line. Test the hypothesis that $\beta = 0.80$.

20. In Exercise 4 find the 95 per cent confidence interval for β.

21. In a study on the gain in weight in grams (y), for different types of rations in 100 calorie units (x), made with samples of ten rats each, the following results were obtained:

Ration A. Line of regression: $y = 40.6 + 0.396x$;

$\Sigma x = 1500$, $\Sigma x^2 = 230,000$

$\Sigma y = 1000$, $\Sigma y^2 = 102,000$

Ration B. Line of regression: $y = 14.1 + 0.677x$;

$\Sigma x = 1120$ $\Sigma x^2 = 128,000$

$\Sigma y = 859$ $\Sigma y^2 = 76,000.$

Test the hypothesis that $\beta_1 = \beta_2$.

CORRELATION

13

"It is evident that the understanding of relations is a major
concern of all men and women."
CASSIUS J. KEYSER
Mole Philosophy

13.1. Introduction

Simple correlation may be defined as "the amount of similarity, in direction and degree, of variations in corresponding pairs of observations of two variables." The principal problem of simple correlation is that of determining the degree of association between these pairs of observations. Thus we may inquire concerning the correlation (degree of association) between the weights and heights of American soldiers, between end-of-year inventories and yearly sales of drug stores, between birth rates and per capita property values of states, and so on.

Regression and correlation are intimately connected as the subsequent sections will demonstrate, but there are essential differences. In what may be termed a pure regression problem, there is an independent or causal variable x, and a dependent variable y. The values of x are assumed to be selected in advance and held fixed; (See Section 12.11) then the corresponding values of y are observed. For example, we may choose a set of varying doses x, of a drug to be administered to groups of animals, and then note the numbers y, succumbing to the dose; or we may select intervals of time at whose end-points x, we observe instrument readings y. In a pure correlation problem we choose a sample of pairs of observations from a bivariate population as, by way of illustration, a company of soldiers whose heights and weights we record, or a group of states of the Union whose birth rates

238

and per capita property values we ascertain. In this latter case we cannot say that the birth rate depends upon the per capita value any more than we can say that the per capita value depends upon the birth rate. Do people have more children when they possess less or do people possess less when they have more children? Here the functional relationship, assuming that such exists, is, from the statistical standpoint, a reversible one.

13.2. Two Lines of Regression

In a correlation problem it is sometimes useful to consider two lines of regression, that of y on x which was treated in the previous chapter, and that of x on y. In the former case we used a least squares line that minimized the sum of the squares of the *vertical* or y-distances of the points from the line; in the latter case we minimize the sum of the squares of the *horizontal* or x-distances of the points from the line.

The problem of finding the equation of this companion line involves nothing new. It requires merely an exchange in the roles of the x- and y-axes. Hence the desired equations may be obtained by replacing x by y and y by x in Equations (12.5), (12.6) and (12.7). The equation of the regression line of x on y will then be:

$$x = a' + b'y \tag{13.1}$$

where

$$b' = \frac{N \Sigma xy - \Sigma x \, \Sigma y}{N \Sigma y^2 - (\Sigma y)^2} \tag{13.2}$$

$$a' = \bar{x} - b'\bar{y}. \tag{13.3}$$

The line (13.1) is more appropriate for estimating an x from a given y.

13.3. The Coefficient of Correlation

Let

$$r^2 = \frac{s_{y'}^2}{s_y^2}, \tag{13.4}$$

the ratio of the explained variance to the total variance; it represents the fraction of the overall variance that is explained by regression. If the points of a dot diagram lie exactly on the regression line, as in Figure 12–7, $s_{y'}^2 = s_y^2$; all of the variance is attributable to regression and $r^2 = 1$. If no regression whatsoever exists, as in Figure 12–6, $s_{y'}^2 = 0$; none of the variability is explained and $r^2 = 0$. The quantity

$$r = \pm \frac{s_{y'}}{s_y}$$

is called the *coefficient of correlation* and is used to measure the degree of association existing between the x's and the y's. Since by (12.22)

$$s_{y'}^2 = s_y^2 - s_e^2$$

$$r^2 = \frac{s_y^2 - s_e^2}{s_y^2} = 1 - \frac{s_e^2}{s_y^2}. \tag{13.5}$$

It is obvious from the preceding discussion that r numerically never exceeds 1. The sign of r is always taken as that of the slope of the regression line. Thus, a positive value of r indicates that y tends to increase with x; a negative value, that y tends to decrease as x increases.

A more practical formula for r is obtained in the following manner.

$$r^2 = \frac{s_{y'}^2}{s_y^2}.$$

$$= \frac{b^2 s_x^2}{s_y^2}, \text{ by (12.28)};$$

then
$$r = \frac{b s_x}{s_y}. \tag{13.6}$$

This formula is useful in converting the regression coefficient b, into the correlation coefficient r, and vice versa. For the line of regression of x on y

$$r = \frac{b' s_y}{s_x}. \tag{13.6a}$$

By means of (12.25) Equation (13.6) becomes

$$r = \frac{N\Sigma xy - \Sigma x\Sigma y}{N^2 s_x s_y}, \tag{13.7}$$

and from the well-known formula for variance

$$r = \frac{N\Sigma xy - \Sigma x\,\Sigma y}{\sqrt{[N\Sigma x^2 - (\Sigma x)^2][N\Sigma y^2 - (\Sigma y)^2]}}. \tag{13.8}$$

This formula involves five basic sums, Σx, Σy, Σx^2, Σy^2, and Σxy, which are readily obtained from an automatic desk calculator. It may also be proved (See Exercise 13.26) that (13.7) can be converted into a formula useful in theoretical work

$$r = \frac{1}{N}\Sigma\left(\frac{x - \bar{x}}{s_x}\right)\left(\frac{y - \bar{y}}{s_y}\right). \tag{13.9}$$

The expression on the right is called the *product-moment* or the *covariance*; r is often termed the *Pearson product-moment coefficient of correlation* after the English statistician *Karl Pearson* (1857–1936) who devised it. In formula (13.9) r is seen to be the arithmetic mean of the products of the corresponding values of the two sets of variates expressed as standard deviates.

13.4. Properties Related to r

From (13.5) one can easily show that

$$s_e^2 = s_y^2(1 - r^2). \tag{13.10}$$

Also for the line of regression of x on y

$$s_e^2 = s_x^2(1 - r^2), \tag{13.10a}$$

and since variances are always positive, this means that r^2 can never exceed 1 and is obviously non-negative.

$$0 \leq r^2 \leq 1.$$

From this it follows that the possible values of r range from -1 to $+1$:

$$-1 \leq r \leq +1.$$

When $r = 0$, also $b = b' = 0$ [See (13.6) and (13.6a)] and the regression lines

$$\left. \begin{array}{l} y = a + bx \\ x = a' + b'y \end{array} \right\}$$

become

$$\left. \begin{array}{l} y = a \\ x = a' \end{array} \right\}.$$

Since the regression lines always pass through the point $(\bar{x}, \bar{y})$ then $a = \bar{y}$ and $a' = \bar{x}$ and the equations become

$$y = \bar{y} \quad \text{and} \quad x = \bar{x}.$$

When $r = \pm 1$, the sum of the squares in both (13.10) and (13.10a) is zero, a minimum; hence, all points must lie on both lines of regression. The only way in which this can happen is to have the lines coincide.

The slopes b and b' depend upon r, as we see by (13.6) and (13.6a). When r is positive, the slope is also, and the lines of regression ascend from left to right [Figures 13–1(a), (b)]. When r is negative, the slope is also, and the lines of regression descend from left to right [Figure 13–1(c)] . When r is numerically very small, the line of regression of y on x is close to the line $y = \bar{y}$. Similarly the line of regression of x on y is close to the line $x = \bar{x}$ [Figure 13–1(a)]. When r is near ± 1, the lines very nearly coincide [Figure 13–1(c)] and when $r = \pm 1$, the lines do coincide [Figure 13–1(d)].

Thus, as r varies from 0 to either $+1$ or -1, the lines of regression rotate from positions on the lines $x = \bar{x}$ and $y = \bar{y}$ toward each other until they coincide. Consequently, the amount of divergence between the two lines of regression gives a simple visual estimate of the degree of correlation.

It can be shown, when standard variables are used, $z_x = \dfrac{x - \bar{x}}{s_x}, z_y = \dfrac{y - \bar{y}}{s_y},$

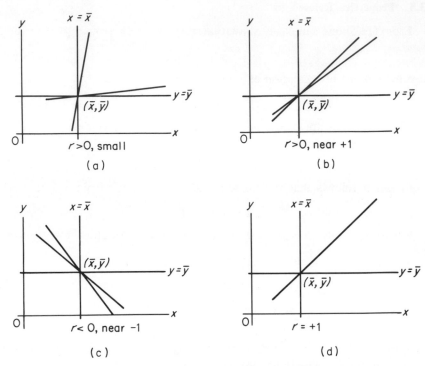

Fig. 13–1.   Positions of regression lines for various values of r.

that the two regression lines always have symmetrical positions with respect to the line bisecting the quadrants through which they pass. This means that the line of regression of z_y on z_x makes the same angle with the positive z_x-axis that the line of regression of z_x on z_y makes with the positive z_y-axis. (Figure 13–2.)

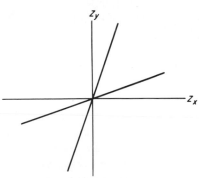

Figure 13–2

From the preceding paragraphs, it appears that the closer the points lie to a line of regression, the more nearly does a simple linear equation express the association between the values of x and y. Thus, a correlation coefficient of nearly 1 (or -1) would seem to indicate a definite relationship between two given sets of values; a coefficient near zero would seem to indicate practically no such relationship.

13.5. The Computation of r

The widespread use of computing machines has diminished but not eliminated the necessity for short-cut methods of pencil calculation such as coding. When coding is unnecessary Formula (13.8) is an excellent one to use. When coding is desirable we let

$$
\left.\begin{aligned}
u &= \frac{x - x_0}{k_x} \\
v &= \frac{y - y_0}{k_y}
\end{aligned}\right\},
\tag{13.11}
$$

so that, from (4.7) and (4.9),

$$
x - \bar{x} = k_x(u - \bar{u})
\tag{13.12}
$$

and

$$
s_x = k_x s_u.
\tag{13.13}
$$

Similarly

$$
y - \bar{y} = k_y(v - \bar{v})
\tag{13.14}
$$

and

$$
s_y = k_y s_v.
\tag{13.15}
$$

Let us take Formula (13.9) and substitute in it the values given in these last four equations.

We obtain

$$
r = \frac{1}{N} \sum \left(\frac{u - \bar{u}}{s_u}\right)\left(\frac{v - \bar{v}}{s_v}\right),
\tag{13.16}
$$

which, [See Formula (13.8)], can be converted into

$$
r = \frac{N\sum uv - \sum u \sum v}{\sqrt{[N\sum u^2 - (\sum u)^2][N\sum v^2 - (\sum v)^2]}}.
\tag{13.17}
$$

The identity of form of Formulas (13.9) and (13.16) as well as that of (13.8) and (13.17) may be expressed by the following theorem:

The coefficient of correlation, r, remains invariant under the transformations (13.11).

13.6. The Computation of r for Ungrouped Data

Let us compute the coefficient of correlation for the birth rates and per capita property values of Table 13–1. The steps in the computation appear in this table. Provisional means $x_0 = 45$ and $y_0 = 3000$ are selected; $k_x = 1$ and $k_y = 100$.

Table 13–1

BIRTH RATES OF NATIVE-BORN WHITES PER 1000 ENUMERATED FEMALE POPULATION, 1920, AND PER CAPITA ESTIMATED VALUE OF ALL PROPERTY, 1922, BY STATES

(Adapted from Raymond Pearl, *The Biology of Population Growth*, Knopf, 1925, p.160.)

COMPUTATION OF r FOR BIRTH RATE AND PER CAPITA PROPERTY VALUE

State	Birth rate x	Per capita value of all property y	$x - 45$ u	$\dfrac{y - 3000}{100}$ v	uv	u^2	v^2
Conn	31	3600	−14	6	−84	196	36
Mass	33	3200	−12	2	−24	144	4
N. Y.	34	3400	−11	4	−44	121	16
D. C.	34	3900	−11	9	−99	121	81
Cal	35	4000	−10	10	−100	100	100
N. H.	37	3100	−8	1	−8	64	1
Vt	39	2400	−6	−6	36	36	36
Ore	40	4200	−5	12	−60	25	144
Ohio	40	3000	−5	0	0	25	0
Wash	41	3600	−4	6	−24	16	36
Me	41	2600	−4	−4	16	16	16
Penn	42	3200	−3	2	−6	9	4
Ind	44	2900	−1	−1	1	1	1
Wis	45	2900	0	−1	0	0	1
Kan	46	3500	1	5	5	1	25
Md	47	2700	2	−3	−6	4	9
Mich	48	2900	3	−1	−3	9	1
Minn	48	3400	3	4	12	9	16
Neb	50	4000	5	10	50	25	100
Ky	55	1500	10	−15	−150	100	225
Va	57	2000	12	−10	−120	144	100
S. C.	59	1400	14	−16	−224	196	256
N. C.	64	1700	19	−13	−247	361	169
Utah	64	3200	19	2	38	361	4
Totals	1074	72,300	−6	3	−1041	2084	1381

$$\Sigma u = -6 \qquad \Sigma u^2 = 2084 \qquad \Sigma uv = -1041$$

$$\Sigma v = 3 \qquad \Sigma v^2 = 1381$$

From Formula (13.17)

$$r = \frac{(24)(-1041) - (-6)(3)}{\sqrt{24(2084) - (-6)^2}\ \sqrt{24(1381) - (3)^2}}$$

$$= -0.614.$$

This sample consists of 24 pairs of variates drawn from a finite population, 48 states plus the District of Columbia (in 1920). There is no large or potentially infinite population from which this sample can be conceived to be drawn. The computed value of r would seem to indicate a fairly high degree of correlation, the negative sign implying that high per capita property values go, in general, with low birth rates. A question somewhat beyond the

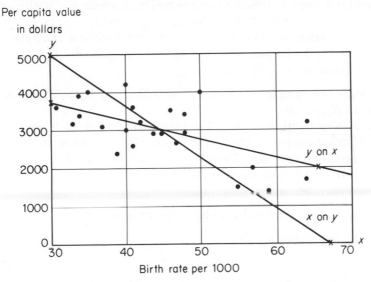

Fig. 13–3. Scatter diagram for birth rates and per capita property value. (Data of Table 13–1.)

scope of this book is the following: Are birth rates and property values directly related to the extent measured by the value, $r = -0.61$, or are both attributes really dependent upon other variables, or latent factors, not revealed by the bare data of Table 13–1? The discussion of this question may be found in the source of this table.

The equations of the lines of regression have been found to be

$$\begin{cases} y = 5252 - 50.0x \\ x = 67.5 - 0.00754y. \end{cases}$$

Here, since we calculated r first, we computed b (-50.0) from (13.6), $b = r\dfrac{s_y}{s_x}$. Similarly $b' = r\dfrac{s_x}{s_y}$. The regression lines are plotted in Figure 13–3, where the crosses indicate convenient points used to determine the lines. A check on the work is made by estimating the coordinates of the point of intersection of the lines. These should be $(\bar{x}, \bar{y})$ or approximately $(45, 3000)$.

13.7. Testing the Hypothesis that $\rho = 0$

For samples of N pairs (x, y) each assumed to be drawn from a normal bivariate population (Section 13.14) with means μ_x and μ_y, standard deviations σ_x and σ_y, and correlation coefficient ρ, the distribution of r has been determined. This distribution has a complicated form but when $\rho = 0$ its equation for a given N defines a curve symmetrical about $\rho = 0$ with a range from -1 to $+1$. See the figure accompanying Table I.

A sample value of r may be most easily tested without computation by means of Table I. In that table the sign of r is ignored. The number of degrees of freedom, $n = N - 2$.

Example. A sample of 15 shows $r = 0.53$. Test $H_0 : \rho = 0$ against the alternative $H_1 : \rho \neq 0$.

From Table I with $n = 15 - 2 = 13$, we find that at the 5 per cent level (two-tail probability), $r = 0.514$, hence r is significantly different from $\rho = 0$ and we cannot accept H_0. Thus for $\alpha = 0.05$ we should concede that a positive correlation exists in the population.

13.8. The Use of the Student-Fisher t-Distribution

For the hypothesis $\rho = 0$,

$$t = r\left(\frac{N-2}{1-r^2}\right)^{\frac{1}{2}} \tag{13.18}$$

is known to have a t-distribution with $N - 2$ degrees of freedom; Hence we may test t with the aid of Table E. Formula (13.18) can be shown to be equivalent to (12.30) with $\beta = 0$.

13.9. The Fisher z-Transformation

When $\rho \neq 0$ the distribution of r is skew and the skewness increases with the numerical value of ρ. (See Figure 13–4.) However, the variable

$$z = \frac{1}{2}\log_e \frac{1+r}{1-r}, \tag{13.19}$$

devised by R. A. Fisher, has approximately a normal distribution which changes little in form with ρ. (See Figure 13–5.)

One does not need to employ logarithms here, for the transformation of r into z is easily effected by means of Table J. Except for very small values of N with ρ near ± 1, we may use z with

$$\sigma_z = \frac{1}{\sqrt{N-3}}. \tag{13.20}$$

Replacing r by ρ we let

$$\zeta = \frac{1}{2}\log_e \frac{1+\rho}{1-\rho}. \tag{13.21}$$

Since the distribution of z is approximately normal, we let

$$z' = \frac{z-\zeta}{\sigma_z} = (z-\zeta)\sqrt{N-3} \tag{13.22}$$

and refer to the normal table D.

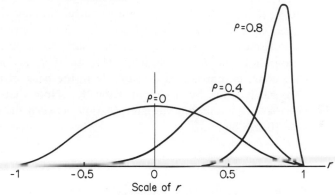

Scale of r

Figure 13–4

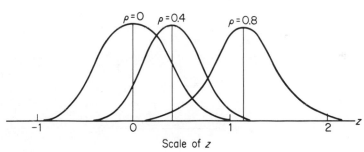

Scale of z

Figure 13–5

EXAMPLE 1.   From a sample of 30 it was found that $r = 0.55$. Is this consistent with the hypothesis that $\rho = 0.40$?

From Table J for $r = 0.55$, $z = 0.62$, and for $\rho = 0.40$, $\zeta = 0.42$. By Formula (13.22)

$$z' = (0.62 - 0.42)\sqrt{27} = 1.04.$$

This deviation of about one sigma is obviously consistent with the hypothesis.

EXAMPLE 2.   Suppose that for $N = 9$, $r = -0.89$. Find the 95 per cent confidence limits for ρ.

From Table J, $z = 1.42$. From the normal table, $P = 0.025$ corresponds to $z' = \pm 1.96$. Then by (13.22), if

$$\pm 1.96 = (1.42 - \zeta)\sqrt{6}$$

$$0.62 < \zeta < 2.22.$$

Converting back to r (Table J) we find that $-0.98 < \rho < -0.55$. Thus there is good evidence that the population correlation coefficient is negative and not less numerically than 0.55.

Given ρ and N, the probability that a sample correlation coefficient does not exceed a given value r, may, for samples of $N = 3$ to 25 principally, be found in *David's* Tables (Reference 19). Useful charts showing confidence belts for ρ corresponding to $P = 0.90, 0.95, 0.98$, and 0.99 are also found there. One such chart is reproduced as Table K in the back of this book.

Example 2 is easily done by means of Table K. Note that the curve corresponding to $N = 9$ is missing but interpolation between the curves for $N = 8$ and $N = 10$ can be used.

13.10. Testing the Hypothesis, $\rho_1 = \rho_2$

Let r_1 and r_2 be two coefficients of correlation obtained from samples of N_1 and N_2 respectively. Let z_1 and z_2 correspond to r_1 and r_2 by virtue of (13.19).

Then by Formula (8.7)

$$\sigma^2_{z_1 - z_2} = \sigma^2_{z_1} + \sigma^2_{z_2}$$

$$= \frac{1}{N_1 - 3} + \frac{1}{N_2 - 3}. \tag{13.23}$$

We may set

$$z' = \frac{(z_1 - z_2) - (\zeta_1 - \zeta_2)}{\sigma_{z_1 - z_2}}$$

which by virtue of the null hypothesis, $\zeta_1 - \zeta_2 = 0$, reduces to

$$z' = \frac{z_1 - z_2}{\left(\dfrac{1}{N_1 - 3} + \dfrac{1}{N_2 - 3} \right)^{1/2}}. \tag{13.24}$$

A theorem in Section 8.5 shows that we may refer z' to a normal scale.

Example. In investigating the correlation between abilities in mathematics and music, a sample of 12 showed $r = 0.30$ while another sample of 19 showed $r = 0.20$. Do these samples indicate a difference between the population correlation coefficients?

$r_1 = 0.30$ and $r_2 = 0.20$ correspond to $z_1 = 0.31$ and $z_2 = 0.20$ respectively. The hypothesis is $\zeta_1 - \zeta_2 = 0$. By (13.24)

$$z' = \frac{0.31 - 0.20}{(\frac{1}{9} + \frac{1}{16})^{\frac{1}{2}}} = 0.26.$$

Since z' is referred to a normal scale, no difference between the population ρ's is indicated.

13.11. The Correlation Table

When the total number of paired values, N, is large, a preliminary analysis of the data may be made by means of a double frequency distribution or correlation table. This consists of a rectangular array of squares or *cells* containing frequencies. The frequency recorded in a given cell represents the number of items belonging to a certain x-class and to a certain y-class simultaneously. In Table 13–2, each cell frequency represents the number of women students having simultaneously a certain weight and a certain height. For example, there were twelve students weighing from 110 to 120 pounds (class mark, 115), and ranging in height from 62.5 to 63.5 inches (class mark, 63). The first step in studying the correlation between height and weight is to construct the correlation table proper, which, in Table 13–2, consists of that portion bounded by heavy lines. If each cell frequency is replaced by an appropriate number of dots, a scatter diagram results. Whether we do this or not (and it is usually unnecessary), it is generally possible to decide whether or not the Pearson coefficient, r, is an appropriate measure of correlation to use. In the example under discussion, it appears that increasing height is associated with fairly steadily increasing weight, so that we may safely assume that we may use "lines" of regression rather than "curves" of of regression. (Section 13.15.)

13.12. The Computation of r from a Correlation Table

Table 13–2 illustrates the method of calculating r when the total frequency, N, is large enough to justify the use of a correlation table. We designate the class marks (mid-values) for the weights by x_i, and for the heights by y_j. The class intervals are 10 (pounds) and 1 (inch), respectively; the provisional means are selected as 125 (pounds) and 64 (inches). Then,

$$\begin{cases} u_i = \dfrac{x_i - 125}{10}, \\ v_i = y_i - 64. \end{cases}$$

The column headed f_y gives the frequencies or *marginal totals* for the heights, y_j. These are obtained by adding the cell frequencies in each row.

Table 13-2

CORRELATION TABLE FOR THE WEIGHTS AND HEIGHTS OF 285 BOSTON UNIVERSITY WOMEN STUDENTS (*Original data*)

Weights in Pounds

$x \rightarrow$ $y \downarrow$	85	95	105	115	125	135	145	155	165	175	185	195	205	f_u	v_i	$f_v v$	$f_v v^2$	$v_i\sum_i u$
57										1				1	−7	−7	49	−35
58														0	−6	0	0	0
59	1	3	4	2										7	−5	−35	175	70
60		3	8	1	5			1						13	−4	−52	208	92
61	1	3	2	4	11	3	2							18	−3	−54	162	54
62			7	8	13	3	1							34	−2	−68	136	48
63			7	12	14	6	6	1			1	1		41	−1	−41	41	8
64		1	8	8	15	10	4	2		1				49	0	0	0	0
65			1	10	10	11	5			1				42	1	42	42	12
66			2	9	4	6	5	2		1				36	2	72	144	22
67				2	2	8	2	3	1				1	24	3	72	216	114
68				1		2	1	1	1					9	4	36	144	48
69						4		1		1				7	5	35	175	65
70					1									1	6	6	36	30
71						1					1			2	7	14	98	42
72														1	8	8	64	8
f_x	2	10	39	57	75	54	26	11	2	5	2	1	1	285		28	1690	578
u_i	−4	−3	−2	−1	0	1	2	3	4	5	6	7	8					
$f_x u$	−8	−30	−78	−57	0	54	52	33	8	25	12	7	8	26				
$f_x u^2$	32	90	156	57	0	54	104	99	32	125	72	49	64	934				
$u_i\sum_i v$	32	75	148	16	0	62	74	51	36	25	42	−7	24	578				

Heights in Inches

$$\Sigma u = 26 \qquad \Sigma u^2 = 934 \qquad \Sigma uv = 578$$

$$\Sigma v = 28 \qquad \Sigma v^2 = 1690$$

From Formula (13.17)

$$r = \frac{(285)(578) - (26)(28)}{\sqrt{285(934) - (26)^2}\ \sqrt{285(1690) - (28)^2}}$$

$$= 0.459.$$

The row labeled f_x yields the frequencies or marginal totals for the weights, x_i. These are obtained by adding the cell frequencies in each column. These two frequency distributions may be used for calculations already familiar. The total frequency, 285, may be obtained by either a horizontal addition of the values, f_x, or a vertical addition of the values, f_y. The two sums together give a valuable check on N. The column headed v_j and the row labeled u_i contain the unit deviations from the provisional means. The next two columns and the next two rows are used to compute the means and the standard deviation, as shown after the table. Only the last column and the last row offer unfamiliar calculations. The last column, headed $v_j \sum_{j} u$, is used to obtain Σuv. Each product in this column is found by multiplying a v_j by the sum of the products of the various u's by the row frequencies corresponding to v_j. Thus, in the fifth row, $j = 5$, $y_5 = 61$; hence, $v_5 - -3$. The sum of the products of the u's in the jth row by their frequencies will be denoted by $\sum_{j} u$, where the subscript j under Σ specifies the sum of the u's in the jth row, each u occurring the number of times indicated by the corresponding frequency.

$$\sum_{5} u = (1 \times -4) + (3 \times -3) + (2 \times -2) + (4 \times -1)$$
$$+ (5 \times 0) + (3 \times 1) = -18.$$

Hence, for this row, $v_5 \sum_{5} u = (-3) \times (-18) = 54$. The sum of all products $v_j \sum_{j} u$ in the last column gives 578. This number is exactly the quantity Σuv, which is theoretically obtained by multiplying a u by its paired v and then multiplying the product uv by its proper cell frequency. However, when products uv for which v is constant (in each row) are selected, the common factor v_j simplifies the arithmetical work considerably.

Similar methods and remarks apply in obtaining the last row, $u_i \sum_{i} v$.

For example, the total, 51, obtained at the foot of the 155-pound column is obtained by multiplying the frequencies in this column by their corresponding v's, adding, and then multiplying the sum by the value $u_8 = 3$. Thus,

$$3[(1 \times -4) + (1 \times -1) + (2 \times 0) + (2 \times 2) + (3 \times 3)$$
$$+ (1 \times 4) + (1 \times 5)] = 51.$$

Since $\Sigma[v_j \sum_j u] = \Sigma[u_i \sum_i v] = 578$, we have again a valuable check on our computation. Formula (13.17) yields $r = 0.459$, as shown after Table 13–2. This value indicates a moderate degree of correlation, so that we may say that the tendency for the weights and heights of women students to increase together seems to be fairly well indicated. This conclusion can be verified to be valid, since the number N in the sample is large, 285. (See Section 13.7.)

The lines of regression can be shown to be the following:

$$y \text{ on } x: \quad y = 56.3 + 0.062x$$

$$x \text{ on } y: \quad x = -92 + 3.40y.$$

In the majority of practical problems, we compute the equation of the regression line first, then, if desired, the correlation coefficient. In data for heights and weights we usually conceive of weight as dependent on height, so that if we wish to estimate a weight for a given height, we should use the second regression equation, x on y, above. For estimating height from weight, we should use the first regression equation.

13.13. Remarks on the Computation of r

(1) *Significant digits.* Data used in correlation problems usually involve not more than three significant digits and generally contain but two. From the hard practical standpoint, figures beyond the second decimal place in the value of r have little significance. A good working rule, therefore, is to use three significant figures in the computation and to round off r to the second decimal place.

(2) *Grouping errors.* No exact rule can be stated concerning the values of N for which data should be grouped into a correlation table. When N is small, say below 100, or better, below 50, the direct method without a correlation table is generally advised. The correlation table used consists of 16 rows and 13 columns. We call this a 16×13 table. It has been shown that as long as a table has dimensions greater than 10×10, the error in r due to grouping of the data into classes is not very large (about 4 per cent). When the table is reduced below the 10×10 requirement, the error in r increases and, in fact, may become so serious as to vitiate the result. Rather than apply cumbersome corrections to correlation tables of less than 10×10 dimensions, it is generally better to compute r without grouping the data, or else to apply, where feasible, other methods as yet not discussed.

Table 13–3 was formed from Table 13–2 by enlarging the class intervals from 10 to 20 pounds and from 1 inch to 3 inches. This gives us a 7×6 table. The value of r is thus changed from 0.46 to a less accurate value, 0.39. (See Exercise 13.25.)

(3) *Calculating machines.* Improvements in the modern computing machine now permit the simultaneous calculation of the quantities Σx, Σy, Σx^2, $2\,\Sigma xy$, and Σy^2 from ungrouped data. Such a procedure eliminates the necessity for making correlation tables and avoids errors due to grouping. (See, for example, P. S. Dwyer, "The Calculation of Correlation Coefficients from Ungrouped Data with Modern Calculating Machines," *Journal of the American Statistical Association*, December, 1940.)

Table 13–3

REDUCED CORRELATION TABLE FOR WEIGHTS AND HEIGHTS
OF WOMEN STUDENTS (*Data of Table* 13–2)

Weights in Pounds

Heights in Inches	90	110	130	150	170	190	210
57					1		
60	8	21	8	1			
63	3	50	57	12		2	
66	1	24	54	19	3		1
69		1	8	5	3		
72			2			1	

There are other methods of measuring correlation which will be mentioned later. (See Sections 13.16, 13.17, and 15.8.) The choice of method depends largely on the value of the total frequency, N, the form of the data, and the objective of the particular statistical investigation.

13.14. The Correlation Surface

The graphical representation of a frequency distribution by means of a histogram or frequency polygon (Section 3.4) and the idealization of the latter into a frequency curve are already familiar. Similar processes are employed in the case of a *bivariate* distribution represented by a correlation table. Let each cell in the table be the base of a solid rectangular column whose height is proportional to the frequency of the cell. The aggregate of columns thus constructed forms a solid histogram, a sort of modernistic building (Figure 13–6). If the dimensions of each cell are k_x and k_y (the class intervals for x and y, respectively), and the total frequency is N, then the volume of this solid histogram will be Nk_xk_y. If the class intervals are reduced to unity, the volume becomes equal to N.

The concept of the solid histogram is not without its practical applications.

For example, a certain shoe store in Boston uses an effective form of it. The lengths of men's shoes from 4 to 12 and the widths from AAA to E constitute a double array of sizes. Upon each cell, for example, that for size $7\frac{1}{2}$ C, is erected a vertical rod upon which uniform washers of constant thickness can be strung. For each pair of shoes sold, a washer is dropped upon the appropriate rod. The number of washers built up on each rod represents the frequency. After a period of time, say a week, the aggregate of cylindrical columns formed by the washers on the rods yields a form of solid histogram that records the distributions of shoe sales by lengths and widths. Orders for new stocks of shoes can be constructed accordingly.

Returning to the discussion of the first paragraph, if we assume the class intervals, k_x and k_y, each to approach zero while the total frequency, N, becomes infinite, in such a way that the product Nk_xk_y remains finite, the rectangular columns will become infinitely slender and infinitely numerous. We assume that their upper bases will approach, as a limiting form, a certain

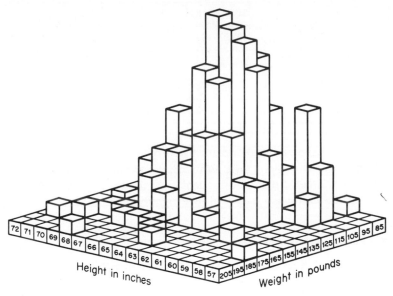

Fig. 13–6. Solid histogram for the Correlation Table 13–2, of heights and weights of 285 Boston University women students.

curved surface called a *frequency surface*. In the case of a so-called *normal bivariate distribution*, this surface will be bell-shaped (Figure 13–7). Any cross section parallel to the (x, y)-plane upon which the array of cells is situated will be an ellipse or circle. The centers of all such ellipses will lie upon the vertical line throughout the "mean point," (μ_x, μ_y); the axes of these ellipses will lie in two vertical planes perpendicular each to each. We shall call these the *axial planes* of the frequency surface. If ρ, the correlation coefficient of the population, is numerically near 1, the ellipses will be slender ones, that is, the *major axis* will be much greater than the *minor axis*. If ρ is numerically near 0, the ellipses will be nearly circular, that is, the major and minor axes will be nearly equal. Any vertical cross section parallel to an axis or to any direction whatsoever will yield a normal curve. The surface will be asymptotic to the base, that is, it will approach infinitely near the base as it extends to infinity in all directions.

The equation of the surface corresponding to a normal bivariate population with means μ_x and μ_y, standard deviations, σ_x and σ_y, and correlation coefficient, ρ, can be shown to be

$$f(x, y) = \frac{1}{2\pi\sigma_x\sigma_y(1 - \rho^2)^{1/2}}\, e^{-\frac{1}{2(1-\rho^2)}\left[\left(\frac{x-\mu_x}{\sigma_x}\right)^2 - 2\rho\left(\frac{x-\mu_x}{\sigma_x}\right)\left(\frac{y-\mu_y}{\sigma_y}\right) + \left(\frac{y-\mu_y}{\sigma_y}\right)^2\right]}. \tag{13.25}$$

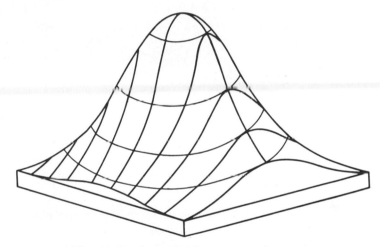

Fig. 13–7. A normal frequency surface.

The detailed study of the normal frequency surface defined by Equation (13.25) must be left to a more advanced course. At this point, however, we call attention to two important aspects of the correlation table and its associated surface. These are best explained by referring to correlation Table 13–2.

In the first place, we observe that the main body of this table may be enclosed in an ellipse whose longer axis slopes diagonally downward from the upper left-hand region of the table, and whose shorter axis slopes diagonally upward. The nearer we approach these axes, the greater, generally, do the frequencies become. These axes lie close to the axes associated with the ideal frequency surface. (Figure 13–7.)

An interesting application of this elliptical distribution is found in gunnery. If a gun is fired at a flat target, the distribution of the shots will exhibit two kinds of errors, one along the direction of fire (longitudinal), the other at right angles to it (latitudinal). Shots that fall short of or exceed the required longitudinal distance will have a greater range of dispersion than shots that fall too far to the right or to the left (latitudinal dispersion). If we imagine each shot to leave a hole in the ground about the target, the aggregate of holes will tend to form an elliptical scatter diagram whose longer axis is in the direction of fire. (Figure 13–8.) We assume here that the gun and its operator are behaving "normally." If the ground about the target is ruled off into squares whose sides are parallel to and perpendicular to the line of fire, the number of shots falling in each square or cell gives a cell frequency.

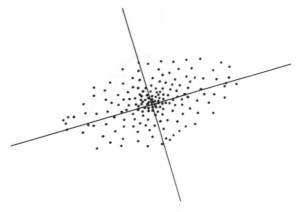

Figure 13–8

Returning to Table 13–2, we may state, in the second place, that it is possible to find, for each column representing a weight class, the average height of students belonging to that class. Similarly, for each row representing a height class, one may compute the average weight of students belonging to that class. For example, the average height of the 54 students belonging to the 135-pound class is 65.1 inches, and the average weight of the 13 students in the 60-inch class is 107.3 pounds. Let us denote the average of the y's, the heights, corresponding to a given weight class, x_i, by the symbol $\bar{y}_i$; and the average of the x's, the weights, corresponding to a given height class, y_j, by the symbol, $\bar{x}_j$. In what follows, the variable subscript i will refer always to the x's, and the variable subscript j to the y's. Let us plot $\bar{y}_i$, against x_i, and then $\bar{x}_j$ against y_j, as in Figure 13–9. The crosses represent the points $(x_i, \bar{y}_i)$, and the circles the points $(\bar{x}_j, y_j)$. It can be shown that the line best fitting the set of crosses is precisely the line of regression of y on x; the line best fitting the set of circles is the line of regression of x on y. For best fit, we weight the means according to the frequencies they represent, and employ vertical distances in the former case and horizontal distances in the latter.

Height in inches

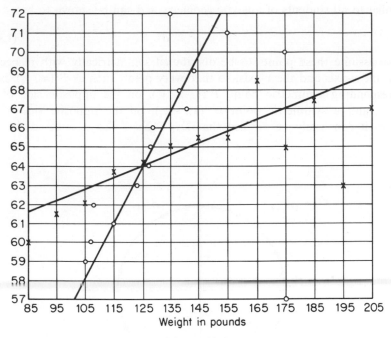

Weight in pounds

Figure 13–9

Passing to the ideal double-frequency distribution as represented by a frequency surface, it can be proved that if the surface is normal, the crosses and circles will lie precisely on straight lines, the *ideal* lines of regression. These facts are important in certain phases of the work on correlation.

13.15. Non-Linear Regression

The theory of correlation as discussed up to this point has been based upon the notion of a straight line best fitting an array of points in a plane. As long as the data given exhibit a fairly linear trend, no serious difficulties ensue. However, if the points seem to lie along a curve, other methods should be used. A bivariate universe that is not normal is represented by a frequency surface that is not normal. It can be shown that the two ideal "lines" of regression associated with nonnormal distributions are not ordinarily both straight lines; one or both may be curved lines. Grossly misleading results may be obtained by computing r upon the hypothesis of a linear trend when the trend is curvilinear.

A simple but striking example of this may be given by considering a number of points lying on the circumference of a circle whose center is the origin and whose radius is a. (Figure 13–10.) The equation of this circle is familiar to all students of analytic geometry and can be shown to be:

$$x^2 + y^2 = a^2. \tag{13.26}$$

If we assume these points to be distributed symmetrically with respect to both the x-axis and the y-axis, so that every point, such as $P:(x_i, y_i)$, has a corresponding symmetric point, $P':(-x_i, y_i)$, on the opposite side of the y-axis, and a symmetric point $P'':(x_i, -y_i)$, on the opposite side of the x-axis, it can easily be verified that $\Sigma x_i = \Sigma y_i = \Sigma x_i y_i = 0$, so that, by

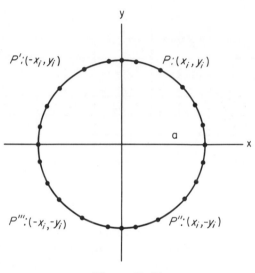

Figure 13–10

(13.7), $r = 0$. This would seem to indicate a complete absence of correlation between the pairs of values (x_i, y_i) corresponding to the points. This result is false, for the correlation is really perfect. The reason is that Equation (13.26) is a mathematical law, a second-degree equation, associating values of x with corresponding values of y. Formula (13.7) involves the assumption that the law connecting x and y is a probable or an approximate law expressed by means of a first-degree equation, the equation of a straight line. The straight lines of regression are easily shown to be the coordinate axes $x = 0$ and $y = 0$, and these fit the points not only equally well but equally badly. It is clear that we should employ a *curve* best fitting the points, that this curve should be a circle, and that this circle fits the points perfectly.

The particular example just discussed is an artificial one, one that would not arise in practical statistical work, but it illustrates the danger of making direct measurements of correlation by standard formulas based on linear trends before the character of the distribution is investigated by means of a scatter diagram or other method. The assumption of a linear trend when a pronouncedly curvilinear trend is present is apt to lead to serious errors in the results sought. The scatter diagram, or its equivalent for large values of N, the correlation table, will often indicate the trends but cannot wholly be relied upon.

13.16. The Correlation Ratio

We have denoted by $\bar{y}_i$ the mean height of students of weight class, x_i, that is, the mean of the heights occurring in the ith column. Let the variance of these heights referred to $\bar{y}_i$ be denoted by $s_{y_i}^2$ and the corresponding total frequency in the ith column by N_i. The mean of the variances, $s_{y_i}^2$, of all the columns, denoted by s_R^2, is found by weighting each variance according to its corresponding total frequency. Hence, by Section 2.6

$$s_R^2 = \frac{1}{N} \sum N_i s_{y_i}^2 \quad \text{where} \quad N = \sum N_i. \tag{13.27}$$

Thus s_R^2 is a measure of the average deviation of the columnar values from their respective means, $\bar{y}_i$.

The mean of the weighted squares of the differences between each column mean and the overall mean will be denoted by s_c^2. Then

$$s_c^2 = \frac{1}{N} \sum N_i (\bar{y}_i - \bar{y})^2. \tag{13.28}$$

As in Section 12.9 it is possible to prove a theorem analogous to that given as (12.22), namely,

$$s_y^2 = s_R^2 + s_c^2. \tag{13.29}$$

Consider the quantity e, defined by the equation:

$$e_{yx}^2 = \frac{s_c^2}{s_y^2} \tag{13.30}$$

$$= 1 - \frac{s_R^2}{s_y^2}. \tag{13.31}$$

When s_R is very small, the dots of the scatter diagram corresponding to the correlation table are closely clustered about their columnar means. In such a case, e_{yx}^2 is very nearly 1. If, on the other hand, s_R is nearly equal to s_y, s_R^2/s_y^2 is nearly unity, e_{yx}^2 is close to zero, and the various $\bar{y}_i$'s must be near $\bar{y}$. Furthermore, s_R can never exceed s_y. This should be apparent from Section 2.11. Hence, e_{yx}^2 ranges in value from 0 to 1, and measures the tendency of the dots to gather into a narrow band, straight or curved, which passes through the $\bar{y}_i$'s. e_{yx} is called the *correlation ratio of y on x*, and is somewhat analogous to the coefficient of correlation, r.

In like manner, we define e_{xy} by means of the equation:

$$e_{xy}^2 = 1 - \frac{s_{R_x}^2}{s_x^2}. \tag{13.32}$$

e_{xy} is the *correlation ratio of x on y*.

An advantage of the correlation ratio, e_{yx}, is that it does not require the use of a best-fitting curve. An alternative measure, the *correlation index*, R_{yx}, makes use of a regression curve of y on x. It is defined by the formula

$$R_{yx}^2 = 1 - \frac{s_e^2}{s_y^2}, \tag{13.33}$$

where s_e is defined as in Formula (12.20), except that the curve replaces the straight line of regression.

Before computing r, the student should be careful to ascertain whether or not he has data which yield linear regression. As has been shown, it is quite possible to obtain a small value of r from data which actually possess high correlation.

We shall not attempt to describe the practical methods of computing correlation ratios and constructing curves of regression, but shall mention the fact that tests for linearity of regression of y on x are based on the difference between r^2 and e_{yx}^2.

13.17. Tetrachoric Correlation

Table 13–4 shows the division of a certain group of men into those physically strong or weak and those having a good standard of living or a poor one. The division of a group into two portions, one of which contains

Table 13–4

Physique \ Standard of living	Good	Poor	Totals
Strong	302	58	360
Weak	347	47	394
Totals	649	105	754

individuals possessing a certain attribute and the other those not possessing it, is called a *dichotomy*. Table 13–4 is dichotomous in two ways: it divides the 754 men into groups physically strong and weak, and into groups having good standards of living and poor ones. We seek a measure of the association between physique and standard of living. Correlation based on such a 2×2 table is said to be *tetrachoric*.

We conceive of men as possessing physiques ranging from very frail to exceedingly robust, and having standards of living varying from abject poverty to great luxury. Although these physical and economic characteristics may be given in a qualitative manner, we imagine them to be normally distributed. In other words, we assume the 754 men to be a sample chosen from a bivariate universe whose correlation surface is a normal one with respect to physique and living standard. This surface is conceived to be divided into four sections by vertical planes at right angles to each other in such a manner that the relative frequencies corresponding to the four portions of this theoretical distribution will be equal, respectively, to the relative frequencies of the four portions of the given table. The fundamental shape of a correlation surface is given by its horizontal elliptical cross sections (Section 13.14), and depends upon the value of r. Our problem, then, in tetrachoric correlation is to find the value of r that yields a correlation surface for which the desired "quartering" is possible. In other words, the r we seek is the r that characterizes a correlation surface exactly fitting the given data. It is unique.

The determination of tetrachoric r, symbolized by r_t, is not always easy, but the construction of useful tables and the employment of certain devices have facilitated the required computation, especially when approximations to r_t are sufficient.

The efficiency of r_t relative to the Pearson r is about 0.40; this means that we would need to have about two and one-half times as many observations in a 2×2 table as in a correlation table in order that r_t be as accurate as r. (See Section 10.11.) Tests of significance for r_t are complicated and will not be discussed here.

In Table 13–4, the value of r_t can be found to be about -0.14, which indicates slight negative correlation. One should also note that any correlation table can be reduced arbitrarily to a 2×2 table. Thus, Table 13–2 may lead to a classification of women students as tall or short, and heavy or light. Students less than 120 pounds in weight might be called light, and those with statures less than 64 inches might be called short. All others would be heavy and tall.

13.18. Multiple Correlation

Simple correlation deals with the degree of association between two variables, such as weight and height. Multiple correlation deals with the degree of interrelationship among three or more variables. Weight may depend not only upon height but upon age as well in the case of growing children. General intelligence in school may be related to grades in mathematics and grades in English. The yield of a potato planting may depend upon the rainfall, the temperature, the amount of fertilizer, and the spacing of the plants.

For simplicity, let us confine our brief discussion to three variables only. Instead of a bivariate distribution, we are considering a *trivariate* distribution. A sample of the former was defined in terms of a two-dimensional array of cells with assigned frequencies. A sample of the latter may be defined in terms of a three-dimensional array of cells, a rectangular parallelepiped subdivided into smaller parallelepipeds with frequencies assigned to many of them. We may visualize such a configuration as an egg-crate, the cells or individual egg receptacles of which lie in horizontal layers, each layer with its cell frequencies corresponding to an ordinary correlation table. The means of the vertical columns of the egg-box determine a three-dimensional configuration of points to which a *regression plane of z on x, y* may be fitted by minimizing vertical distances. The equation of this plane, of the first degree in x, y, and z, may be shown to have the form

$$z = a + bx + cy, \tag{13.34}$$

where a, b and c may be evaluated in terms of the standard deviations s_x, s_y, and s_z and the ordinary coefficients of correlation, r_{xy}, r_{xz}, r_{yz}, where the subscripts indicate the pairs of variables involved.

Corresponding to a given pair of values (x, y) there will be always two values of z: an actual value, z, given by the data, and an estimated value, z', obtained by substituting the given values (x, y) in Equation (13.34). Then the correlation coefficient for the pairs of values of z and z' is called the *multiple correlation coefficient* of z on xy and is denoted by $r_{z.xy}$. Thus,

$$r_{z.xy} = \frac{\dfrac{1}{N}\sum(z - \bar{z})(z' - \bar{z}')}{s_z s_{z'}}.$$

In a similar manner, we may define multiple correlation coefficients of x on yz and of y on xz, symbolized by $r_{x.yz}$ and $r_{y.xz}$, respectively. These may all be computed in terms of the simple correlation coefficients r_{xy}, r_{xz}, and r_{yz}.

For further discussions of the general subject of multiple correlation involving not only three but more than three variables, the reader should consult other works. (See, for example, References 1, 9, and 17.)

13.19. Partial Correlation

In the case of a three-way distribution, one of the "egg-crate" type, it is, of course, possible to study the association between x and y for a given value of z, that is to say, the association existing between values of x and y in any given layer of the egg-box or, as it is sometimes called, a slice or a slab of the trivariate distribution. For example, we may measure the correlation between the heights and weights of American schoolboys of age 16, or the correlation between grades in mathematics and grades in English for high-school boys having intelligence quotients of 110.

The symbol $r_{xy.z}$ represents the *partial correlation coefficient* between x and y for a fixed value of z. There are many cases where we wish to study the degree of association between two variables for various values of a third variable.

13.20. The Uses and Misuses of Correlation

The general subject of correlation is a vast one, and one that cannot be comprehensively treated here. Nevertheless, enough has been said about simple, multiple, and partial correlation to enable the critical student to avoid some of the pitfalls besetting one who attempts to employ correlation methods and to interpret the results. It should be useful, at this point, to make certain remarks and to summarize certain facts concerning the general problem of measuring degree of association.

In the first place, there should be some intelligent basis for proceeding with a proposed study of correlation. It is possible to pair the most disparate pair of variables and grind out corresponding useless values of r. The height of the morning tide and the number of shares sold at the New York Stock Exchange would form an extreme example.

In the second place, if we are proceeding with an investigation of simple correlation, we should be reasonably sure that the suspected association exists essentially by itself and is not a result of some other latent factor or factors. Any possible correlation between school ability in, say, Latin and mathematics, might be due more fundamentally to the degrees of general native intelligence of the students than to fancied common elements in the two disciplines. If other known factors are plausibly involved, it is safer, perhaps, to compute partial correlation coefficients in order to see how much, if any, contribution is made by other factors.

13.21. Multiple Linear Regression

Intimately related to multiple and partial correlation is the concept of multiple regression. We shall confine our remarks to the simplest type, multiple *linear* regression, that is defined by an equation of the first degree in the variables. As an example, the equation of the regression plane (13.34) is regarded as an equation for estimating z corresponding to given values of x and y. The regression coefficients b and c specify the relative contributions of x and y respectively, to the estimate of z.

Multiple regression equations are generally more useful than multiple correlation coefficients. The labor of calculating the regression coefficients for a large number of variables is expedited by the systematic plan of computation and checking known as the Doolittle method.

13.22. Rank Correlation

Sections 15.6 and 15.8 explain the use of ranks in testing for independence of two variables. The advantage of this method is the fact that no assumptions concerning the form of the basic frequency functions, such as normality, are required.

EXERCISES

Compute r by means of Formula (13.8) for the following:

1. The pairs of values in Exercise 12.1.

2. The test scores in Exercise 12.2.

3. The pairs of values in Exercise 12.4.

4. The calculus scores in Exercise 12.7.

5. Convert any slope, b, that you found in the preceding exercises into r by means of Formula (13.6). Use your result to check your answer in Exercises 1–4.

6. Compute r by means of Formula (13.17).

$$x: \quad 20 \quad 50 \quad 70 \quad 40 \quad 10$$
$$y: \quad 400 \quad 200 \quad 100 \quad 200 \quad 500$$

7. The average daily prescription sales, x, and the average daily sales of all other items, y, in 15 drug stores in 1959 showed the following results (in rounded figures):

$x:$ 300   220   330   190   280   110   240   310   120   155   250

$y:$ 900   600   550   300   420   330   900   1390   400   500   500

$x:$ 335   270   164   193

$y:$ 1100   1100   200   850

Compute r by means of the coding formula (13.17).

8. From the dot diagrams below estimate the value of r associated with each.

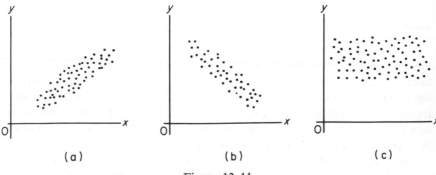

(a)   (b)   (c)

Figure 13–11

9. From the graphs of the regression lines shown below estimate the value of r in each case.

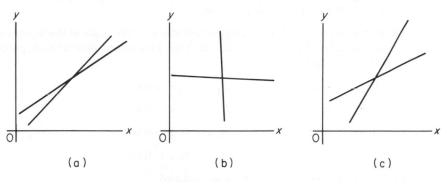

(a)   (b)   (c)

Figure 13–12

10. Examine the scatter diagram shown in the figure below. What is wrong with the following statements? (a) $r = 0.95$. (b) The line of regression of y on x has the equation $y = 11 + 2x$.

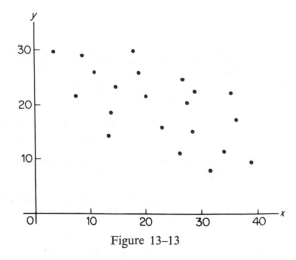

Figure 13–13

11. For 563 students majoring in Literature the correlation coefficient for verbal scores and literature scores on a Graduate Record Examination was 0.59. For 343 students majoring in Chemistry the correlation coefficient for verbal scores and chemistry was 0.43. Interpret each of these statements.

12. Five specimens of castings showed a correlation coefficient of 0.95 for hardness and tensile strength. (a) Test $H_0:\rho = 0$ against the alternative, $H_1:\rho \neq 0$ with $\alpha = 0.05$. (b) Test $H_0:\rho = 0.60$ against $H_1:\rho > 0.60$ with $\alpha = 0.02$.

13. In a study of the relation between the average daily prescription sales and other sales, 19 drug stores showed $r = 0.45$. (a) Make three different statements about the information that this value gives you. (b) Test $H_0: \rho = 0.40$ against $H_1: \rho > 0.40$.

14. By means of Table I test the hypothesis that $\rho = 0$ for each of the following sample correlation coefficients, with Type I errors and alternative hypotheses as indicated.

   (a) $r = \quad 0.29, N = 42;\qquad H_1: \rho \neq 0, \alpha = 0.05.$

   (b) $r = \quad 0.67, N = 18;\qquad H_1: \rho > 0, \alpha = 0.01.$

   (c) $r = -0.44, N = 27;\qquad H_1: \rho < 0, \alpha = 0.025.$

   (d) $r = -0.87, N = 10;\qquad H_1: \rho \neq 0, \alpha = 0.05.$

15. Find confidence intervals for ρ as indicated.

   (a) $r = \quad 0.48, N = 16, 95\%.$

   (b) $r = -0.65, N = 32, 98\%.$

   (c) $r = \quad 0.91, N = 8, 90\%.$

   (d) $r = -0.32, N = 23, 95\%.$

   (e) $r = \quad 0.21, N = 75, 96\%.$

16. A person claims that a value of $r = -0.50$ from a sample of 20 indicates moderately good correlation (say $\rho < -0.40$) in the population from which the sample was drawn. Do you agree? To answer this question find the 95 per cent confidence interval for ρ.

17. In investigating the effects of a certain drug on a group of 19 patients, it was found that $r = 0.43$. Find 95 per cent confidence limits for ρ.

18. Test the significance of the value, $r = 0.44$ from a sample of 68 when it is assumed that $\rho = 0.50$.

19. Which is more significant, a value of $r = 0.80$ from a sample of 10, or a value of $r = 0.30$ from a sample of 100?

20. For 78 X College grades and B Medical School grades, $r = 0.65$; for 93 Y College grades and B Medical School grades, $r = 0.58$. Is there indication here that X College prepares its students better than Y College does for the B Medical School?

21. Of two samples, the first, of 20 pairs, yields a value, $r = 0.60$; the second, of 25 pairs, yields $r = 0.80$. Are these two values of r significantly different?

Compute the value of r for each of the correlation tables given in Exercises 22 and 23, and interpret the result. Find the equations of the lines of regression as directed by your instructor.

22. Grades of medical students in Anatomy (x) and Biology (y). Numerical equivalents for the letter grades are indicated. (Original data.)

y \ x		A	A−	B+	B	B−	C+	C	C−	D+	D	D−	F
		5	4	3	2	1	0	−1	−2	−3	−4	−5	−6
A	5	1		2	10	5	3	6		1		1	
A−	4	1	1	4	5	6	7	5	3	1			
B+	3	3	4		12	14	12	7	2		3		1
B	2	3	1	11	36	20	31	28	10	2	4		2
B−	1	2	2	5	6	11	30	18	9	1	4		2
C+	0	2	2	3	8	10	24	26	16	1	7	1	2
C	−1	2	1	2	13	18	17	30	10	1	7	3	3
C−	−2			1	4	2	3	9	6	2	3	2	2
D+	−3			1		2	2	4	1		1		
D	−4					1	1	2	1	1	2		

23. Verbal aptitude scores (x) and mathematical aptitude scores (y). Original data.)

y \ x	340	380	420	460	500	540	580	620	660	700	740
720							1			1	
680				1	1						
640				2	2			2	1		1
600				2	1	1	4	2			
560	1	1	1	7	3	4	4	5	2	1	
520	3	2	4	6	4	7	4	1	3	1	
480			5	3	7	4	3	3			
440		2	3	6	9	4	4	3	1	1	
400	3	1	1	6	5	3	1	1	1		
360			1	1	1	2		1			
320	1	1	2	1	1		1				

24. The following correlation coefficients have been obtained from actual data. Assuming the samples involved to be large, how would you interpret each result?

(*a*) The weights and lengths of babies: $r = 0.62$ to 0.64.

(*b*) Right and left first joint of the ring finger: $r = 0.93$.

(*c*) Strength of pull and stature: $r = 0.22$ to 0.30.

(*d*) Number of children and per cent of desertions: $r = -0.92$.

25. Compute r for the data of Table 13-3 and verify the result given in Section 13.13.

26. Prove that Formula (13.9) is equivalent to (13.7).

THE ANALYSIS OF
VARIANCE

14

"I have not kept my square, but that to come shall all be done
by the rule."
Antony and Cleopatra
Act II, Scene 3

14.1. Simple Analysis of Variance

From the relations existing among sums of squares, Fisher and others have developed a powerful method of analyzing measurement data when these have been appropriately classified. This method, for the simplest case, will be introduced by means of an example.

Consider the data of Table 14–1, where the yields in pounds of lima beans on 20 plots of ground subjected to four different treatments, five plots per treatment, are given within the heavier ruled lines. For the present we

Table 14–1
TREATMENTS

Plots	1	2	3	4	Totals
1	26.3	18.5	36.9	39.8	
2	30.0	21.1	21.8	28.7	
3	54.2	29.3	24.0	21.2	
4	25.7	17.2	18.5	39.4	
5	52.4	12.4	10.2	29.0	
Σx	188.6	98.5	111.4	158.1	556.60
Σx^2	7935.58	2095.55	2859.14	5250.53	18,140.80
$(\Sigma x)^2$	35,569.96	9702.25	12,409.96	24,995.61	82,677.78

$$(\Sigma x)^2 = (556.60)^2 = 309,803.56.$$

269

shall disregard any possible variation among the yields of the plots themselves due to differences of soil, location, and to other possible factors.

Let us assume that the 20 yields constitute a sample drawn from a normal population with variance σ^2. Then each of the four subsamples consisting of five plot-yields per treatment is also a sample drawn from the same population. There are three sums of squares that we shall presently find useful. They are obtained from the sums found at the bottom of the table.

(1) *The "over-all" sum of squares or total variation, V.* By Formula (2.7) applied to the entire sample,

$$s^2 = \frac{1}{N} \sum (x - \bar{x})^2$$

$$= \frac{1}{N} \sum x^2 - \left(\frac{1}{N} \sum x \right)^2,$$

whence
$$Ns^2 = \sum x^2 - \frac{1}{N} (\sum x)^2. \tag{14.1}$$

The quantity $V = Ns^2 = \overset{N}{\sum} (x - \bar{x})^2$ is an over-all sum of squares. From Table 14–1, V is best computed with the aid of (14.1).

$$V = 7935.58 + 2095.55 + 2859.14 + 5250.53 - \frac{309,803.56}{20}$$

$$= 18,140.80 - 15,490.68 = 2650.12.$$

The quantity $\frac{1}{N} (\sum x)^2$ is sometimes called a *correction term*.

It will be necessary to distinguish clearly by an appropriate notation among sums taken column-wise, row-wise, and "over-all-wise." If a table has N values arranged in r rows and k columns, then

$$\sum_{j=1}^{k} \sum_{i=1}^{r} x_{ij}, \quad \text{or simply} \quad \overset{N}{\sum} x,$$

will indicate their complete sum. From the table

$$\overset{N}{\sum} x = 188.6 + 98.5 + 111.4 + 158.1 = 556.60.$$

(2) *The "between-column" sum of squares or variation, V_c.* Let $\bar{x}_j$ be the mean of the jth column. Then

$$\bar{x}_j = \frac{1}{r} \sum_{i=1}^{r} x_{ij}. \tag{14.2}$$

For k column means, Formula (14.1) becomes

$$ks_{\bar{x}}^2 = \sum_{j=1}^{k} \bar{x}_j^2 - \frac{1}{k}\left(\sum_{j=1}^{k} \bar{x}_j\right)^2$$

$$= \sum_{j=1}^{k}\left[\frac{1}{r}\sum_{i=1}^{r} x_{ij}\right]^2 - \frac{1}{k}\left[\sum_{j=1}^{k}\frac{1}{r}\sum_{i=1}^{r} x_{ij}\right]^2$$

$$= \frac{1}{r^2}\sum_{j=1}^{k}\left[\sum_{i=1}^{r} x_{ij}\right]^2 - \frac{1}{kr^2}\left[\sum_{j=1}^{k}\sum_{i=1}^{r} x_{ij}\right]^2.$$

If we define

$$V_c = r\sum_{j=1}^{k}(\bar{x}_j - \bar{x})^2 = rks_{\bar{x}}^2, \tag{14.3}$$

then

$$V_c = \frac{1}{r}\sum_{j=1}^{k}\left(\sum_{i=1}^{r} x_{ij}\right)^2 - \frac{1}{N}\left(\sum^{N} x\right)^2. \tag{14.4}$$

The last formula may be remembered as follows:

The between-column sum of squares is $\frac{1}{r}$-th the sum of the squares of the column-sums minus the correction term.

Referring to the sums at the bottom of Table 14–1, we find by applying (14.4) that

$$V_c = \tfrac{1}{5}\,(35,569.96 + 9,702.25 + 12,409.96 + 24,995.61) - \frac{309,803.56}{20}$$

$$= \tfrac{1}{5}\,(82,677.78) - 15,490.68 = 1044.88.$$

The reason for the use of this quantity may be explained as follows. If the samples were really drawn from the same population, that is, if all treatments had identical effects on the yield, then the means of the samples would tend to be near the population mean. If, however, the treatments do produce unusually different yields, the means would tend to vary considerably from the general mean and thus increase the variation, V_c. See (14.3). We shall be testing for such an increased sum of squares. The use of V_c, the between-mean variation, stems from the relation

$$\sigma_{\bar{x}}^2 = \frac{\sigma_x^2}{N},$$

with the aid of which we estimate the population variance, σ_x^2, from the population variance of the mean, $\sigma_{\bar{x}}^2$:

$$\sigma_x^2 = N\sigma_{\bar{x}}^2.$$

(3) *The within-column sum of squares or residual,* **R**. Let s_j^2 be the variance for the *j*th column. Then the sum of the squares for the *j*th column is, by (14.1),

$$rs_j^2 = \sum_{i=1}^{r} x_{ij}^2 - \frac{1}{r}\left(\sum_{i=1}^{r} x_{ij}\right)^2.$$

For example, for the first column of the table

$$5s_1^2 = 7935.58 - \tfrac{1}{5}\,(35{,}569.96)^2$$

$$= 821.59.$$

The residual sum of squares is obtained by adding all the within-column sums of squares.

$$R = \sum_{j=1}^{k} rs_j^2. \tag{14.5}$$

The computation just suggested by (14.5) is given for expository reasons. A better method arises from the formula

$$R = \sum^{N} x^2 - \frac{1}{r}\sum_{j=1}^{k}\left(\sum_{i=1}^{r} x_{ij}\right)^2, \tag{14.6}$$

where the second term on the right is seen to be the same as the first on the right of (14.4). Thus,

$$R = 18{,}140.80 - \tfrac{1}{5}\,(82{,}677.78) = 1605.24.$$

If we recall Theorem 2-6 of Section 2.11, we know that for each sample the smallest sum of squares arises when it is taken about its own sample mean. The residual, *R*, will not be affected by the variation of the means among the samples. Thus, *R* will be useful in estimating the variability due to chance effects only, and not that due to possible treatment differences.

Table 14–2

Source of variation	Sum of squares	Degrees of freedom	Estimated variance
Plots (within samples)	$R = 1605.25$	16	100.33
Treatments (between samples)	$V_c = 1044.88$	3	348.29
Total	$V = 2650.12$	19	

The three different sums of squares obtained are placed in the second column of Table 14–2, which is an example of a typical *analysis of variance table*. A basic algebraic fact, proved by adding (14.4) and (14.6), is that the first two sums of squares make up the total sum of squares. That is

$$V = V_c + R. \tag{14.7}$$

We say, then, that the total variation has two components, a between-sample variation and a residual variation. An unbiased estimate of the residual variance, σ_R^2, is obtained by pooling the within-sample variances according to Formula (10.8).

$$\hat{\sigma}_R^2 = \frac{\sum_{j=1}^{k} r s_j^2}{N - k}. \tag{14.8}$$

If the treatments really produce significantly different yields, the unbiased estimate of the population variance obtained from the means of the sample will be significantly larger than the one just discussed. Since there are four means, the number of degrees of freedom for $\hat{\sigma}_{\bar{x}}^2$ is three. By Formula (10.5), applied to a sample of r,

$$\hat{\sigma}_{\bar{x}}^2 = \frac{\hat{\sigma}_x^2}{r}$$

hence

$$\hat{\sigma}_x^2 = r\hat{\sigma}_{\bar{x}}^2,$$

and by (10.3)

$$\hat{\sigma}_x^2 = \frac{1}{k - 1} \cdot r \sum_{j=1}^{k} (\bar{x}_j - \bar{x})^2$$

$$= \frac{V_c}{k - 1}.$$

It is possible to prove that V_c and R are independent, hence we may employ the F-test to see if the between-sample estimate is significantly larger than the residual estimate. The number of degrees of freedom for the latter is $N - k$. The third column of Table 14–2 illustrates the fact that the sum of the numbers of degrees of freedom for the two independent sums of squares equals the number of degrees of freedom for the total. Each sum of squares divided by its corresponding number of degrees of freedom yields an estimated variance (fourth column).

$$F = \frac{V_c}{k - 1} \bigg/ \frac{R}{N - k}$$

$$= {}^{348}\!/_{100} = 3.48,$$

with $n_1 = 3$, and $n_2 = 16$. Because we are looking for excessive variation among the means, we use only the right tail of the F-distribution. Since $F_{0.05} = 3.24$, this value of F is just significant at the 5 per cent level. There is then statistical evidence for genuine treatment effects.

14.2. Remarks

(a) In some data there is high variability within the samples and smaller variability between the means, that is, despite relatively large chance effects the sample averages remain fairly stable. In such a case F will be less than one.

(b) The justification for the saving of so many digits in the computation lies mainly in the fact that calculations of this kind are ordinarily done on a computing machine where rounding off is inconvenient and more accurate checking is possible.

(c) Although the quantity $V/N - 1$ provides an estimate of the over-all variance, it is of no use in an F-test. For this reason the last line of Table 14–2 is not completed.

(d) The analysis of variance procedure just illustrated can be extended to the case where the samples are of unequal size. Thus the fixed subsample size, r, is replaced by r_j where $j = 1, 2, ..., k$. Formula (14.4) then becomes

$$V_c = \sum_{j=1}^{k} \left[\frac{\left(\sum_{i=1}^{r_j} x_{ij} \right)^2}{r_j} \right] - \frac{1}{N} \left(\sum^{N} x \right)^2. \tag{14.9}$$

Thus each squared column-sum is divided by its sample number and then the sum of these is taken. Of course

$$N = \sum_{j=1}^{k} r_j.$$

EXERCISES

1. The scores of 4 groups of 5 pupils each are shown in the table below. Do the group means exhibit unusual differences?

GROUPS

	A	B	C	D
1	110	99	81	87
2	94	97	101	91
3	116	90	87	98
4	90	109	107	103
5	105	115	109	91

2. The heights in inches of 3 different racial groups of 6 each follow. Is there evidence of stature variation among these groups?

	Race A	Race B	Race C
1	68	62	69
2	64	69	72
3	70	69	67
4	64	60	68
5	65	63	62
6	71	67	70

3. Three groups of 4 rats each were injected with commercial Intocostrin and the number of minutes that elapsed before a reaction took place were recorded with the following results. What conclusions may be drawn?

	A	B	C
1	11	12	2
2	8	10	5
3	7	17	2
4	6	7	7

4. Given the data shown in the table below. Does the hour of the day influence the per cent of defectives?

No. of sample \ Hours	1	2	3	4	5	6	7	8
1	2.0	4.2	1.2	2.3	4.2	0.6	2.2	1.7
2	2.3	3.8	2.1	1.9	2.0	2.6	1.9	0.4
3	2.1	3.6	2.0	1.0	4.0	1.4	3.2	0.8
4	2.5	3.6	2.8	3.9	3.5	1.3	3.5	0.0
5	2.6	3.8	2.4	2.2	3.3	1.5	3.9	0.5

NONPARAMETRIC STATISTICS

15

"The mark of maturity is the capacity for reflective commitment.
This involves, on the one hand, full recognition of all the funded
wisdom of the race, full respect for all the knowledge available
to man and relevant to any specific decision. It also involves the
recognition that knowledge is never complete, that all the
evidence is never in, and that we must, of necessity, decide and
act on the basis of partial information and take the risk of being
partially wrong."

THEODORE M. GREENE

*From an address, "Formulating a Philosophy of Life," given at the
Buck Hill Falls Conference on Preprofessional Education*
November 27, 1950

15.1. Introduction

In most of the statistical situations discussed up to this point we assumed
or knew that the form of the distribution was normal, or binomial, or that of
Poisson, and so on. It was then possible to test hypotheses concerning the
parameters, μ, σ, p, $\mu_x - \mu_y$, or to make estimates of them. Sometimes,
however, we have no prior knowledge of the distribution of the data under
investigation, and attempts to guess its form often result in serious error.

The methods to be discussed in this chapter are called *nonparametric* or
distribution-free methods and make no assumptions about the underlying
distribution function other than that of continuity. Chi-square tests of
independence and goodness of fit (Chapter 11) are examples of nonparametric
techniques. The great advantage of a distribution-free method is clear from
its name. Other advantages appear in the relatively simple calculations
required and in the applicability to data not available as exact measurements
but only in qualitative or relative classifications.

The counting of categories (for example, plus and minus signs), the arrangement of observations in order of magnitude, the assignment of ranks to such an order, the use of the median instead of the mean, of the quartile deviation instead of the standard deviation—these are some of the devices

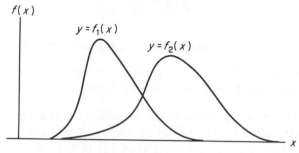

Figure 15–1

used to circumvent assumptions concerning the nature of the basic distribution. The explored area of nonparametric statistics is still relatively small but is expanding at a very rapid rate. Only a limited number of useful methods will be shown in this chapter.

When dealing with two samples nonparametrically, we imagine that they might have come from two populations, $y = f_1(x)$ and $y = f_2(x)$, different in "location" along the x-axis, or different in form, or both. Generally it is the first difference that is of paramount interest. One method of discovering a possible difference in location between the two curves (Figure 15–1) is by means of the medians; another method is by the predominance of higher values in one sample than in the other.

15.2. The Tchebycheff Inequality

An important relation discovered by the French mathematician, Bienaymé (1796–1878), and later by the Russian mathematician, Tchebycheff (1821–1894), is important because it does not depend upon the form of the distribution. The proof is brief.

Let x be a discrete chance variable that assumes the mutually exclusive values $x_1, x_2, ..., x_N$, with probabilities $p_1, p_2, ..., p_N$, respectively, where $\sum_{i=1}^{N} p_i = 1$. By Section 5.7

$$\sum_{i=1}^{N} p_i(x_i - \mu)^2 = \sigma^2. \tag{15.1}$$

Let x_i' represent the values of x_i for which $|x_i - \mu| \geq k\sigma$, where $k > 1$. Then for p_i' corresponding to x_i'

$$\sum p_i'(x_i' - \mu)^2 \leq \sigma^2.$$

Since $(x_i' - \mu)^2 \geqq k^2\sigma^2$, it follows that

$$k^2\sigma^2 \sum p_i' \leqq \sigma^2$$

whence

$$\sum p_i' \leqq \frac{1}{k^2}.$$

Thus we have the following theorem bearing the name of Tchebycheff:

The fraction of observations in a population deviating numerically from the mean μ, *by more than* k *times the standard deviation* σ, *cannot exceed* $1/k^2$.

For example, in any population the probability that a random variable deviates from the mean by as much as 5σ is less than or equal to $\frac{1}{25}$. Stated differently, we may say that not more than $\frac{1}{25}$ of the variates can deviate from the mean by as much as 5σ.

The proof above presupposes that the variable x is discrete. Students familiar with integral calculus should find it instructive to replace the summations above by integrals and carry through the proof for the continuous case. Thus the sum $\sum_{i=1}^{N} p_i = 1$ corresponds to $\int_{-\infty}^{\infty} f(x)\,dx$ where $f(x)$ is the frequency or density function. Equality (15.1) becomes

$$\int_{-\infty}^{\infty} f(x)(x - \mu)^2\,dx = \sigma^2.$$

15.3. The Law of Large Numbers

If we apply Tchebycheff's inequality to the distribution of the relative number of successes, x/N, in N binomial trials with probability p, we may say that the probability that x/N deviates from $\mu_p = p$ by more than $k\sigma$ or its equivalent $k\sqrt{pq/N}$ is less than $1/k^2$. If we let $d = k\sqrt{pq/N}$ so that $k = d\sqrt{N/pq}$ this probability is less than pq/d^2N. As N becomes infinite this maximum probability approaches zero. This gives us the so-called (weak)

Law of large numbers. In N *trials the probability that the relative number of successes* x/N, *deviates numerically from* p *by more than* d $(d > 0)$, *approaches zero as* N *becomes infinite.*

The law for a continuous variable x, is proved in like manner and should be an easy exercise for the student.

The probability that the mean $\bar{x}$, *of a sample of* N *deviates from the population mean* μ, *by more than* d $(d > 0)$ *approaches zero as* N *becomes infinite.*

Note that the laws above hold for any positive d, no matter how small it may be.

15.4. The Sign Test

Consider N pairs of observations (x_i, y_i), $i = 1, 2, ..., N$, and their corresponding differences $d_i = x_i - y_i$, obtained under the conditions described in Section 8.11. If we delete any pair of observations for which $d_i = 0$, and change N to the number of d's remaining, then the sign of d_i will be either positive or negative. We may test the hypothesis that the probability p, of getting a plus sign equals the probability of getting a minus sign. Let x be the number of occurrences of the less frequent sign. Then for $p = \frac{1}{2}$ we have the binomial probability

$$P(x) = \frac{N!}{x!\,(N-x)!}\,(\tfrac{1}{2})^x(\tfrac{1}{2})^{N-x}$$

of obtaining the less frequent sign just x times in N trials or a probability

$$\sum_{x=0}^{r} \frac{N!}{x!\,(N-x)!}\,(\tfrac{1}{2})^N \tag{15.2}$$

of obtaining that sign not more than r times.

Consider the data of Exercise 8.38. The 15 d's in order, are

$$-2, -7, -1, -3, -8, -3, -6, +5, -6, 0, -5, -8, -2, -6, -5.$$

Eliminating the zero difference we have 14 signs of which only one is positive, thus $N = 14$ and $r = 1$. Then the probability for zero or one plus sign is

$$(\tfrac{1}{2})^{14} + \frac{14!}{13!\,1!}\,(\tfrac{1}{2})^{14} = 0.000916.$$

If the expectation is, that if the two instruments do yield different readings, sphygmomanometer C will read generally lower, then the one-tail probability just found, emphatically refutes the hypothesis of a zero median in favor of a negative median; that is, instrument C is judged to read lower than W. Otherwise the two-tail probability 2×0.000916 is calculated. This refutes the hypothesis of a zero median in favor of a nonzero median.

Remarks on the sign test.   (1) The t-test for paired observations is based on the assumption of normality (Section 8.11) and if this assumption is satisfied, is more efficient than the sign test. The normality of d is generally assumed to exist if the conditions under which the pairs of readings were made are essentially identical. However, if, for example, the readings were made sometimes on the left arm and sometimes on the right, sometimes in standing, sitting, and reclining positions, or under other varying conditions, the t-test would not be valid.

(2) The t-test for paired observations shows $t = 4.21$ for 14 degrees of freedom. For a two-tail test $t_{0.01} = 2.98$. In fact, from more extensive

tables of t (Reference 25), $P(t \geq 4.21) = 0.00044$. Clearly the hypothesis, $\mu_d = 0$ is decisively rejected. Since the sign test is easily made and, in this example, produces a very small probability, it is unnecessary to carry out a t-test.

(3) Table L eliminates the necessity for the tedious calculation of probabilities. For our example where $N = 14$ and $r = 1$, if we desire a two-tail test, we find that for $\alpha = 0.05$, r must be at least as small as 2. Since our $r = 1$, we reject the hypothesis of a zero median. For a one-tail test we halve α.

(4) Often an answer to a question like the following is desired: Does sphygmomanometer W read higher than C by an average of 4 millimeters of mercury? In such a case d_i becomes $x_i - (y_i - 4)$ and a two-tail test is used for the alternative that the average may be more or less than 4.

15.5. The Run Test for Randomness

Example. Suppose that a chemical product is tested every half hour for the per cent present of a certain ingredient. Let the per cents recorded at successive half-hour intervals for one day be as follows:

$$12.1 \quad 13.2 \quad 15.4 \quad 15.9 \quad 12.6 \quad 14.7 \quad 11.5 \quad 12.4$$
$$12.3 \quad 16.4 \quad 12.0 \quad 13.0 \quad 16.9 \quad 13.0 \quad 15.0 \quad 15.4$$

The median for these variates is 13.1. We replace observations below the median by a minus sign and those above by a plus sign to obtain the following sequence:

$$- + + + - + - - - - + - - + - + +$$

We state that this sequence contains 10 runs of the following numbers in each: 1, 3, 1, 1, 3, 1, 2, 1, 1, 2. If the fluctuations in the per cents were random, such as might arise by chance, we should not expect on the average, too few or too many runs among those possible. Thus we may test the hypothesis, H_0, that the per cents belong to a single population.

Let u equal the number of runs in the 16 signs above; then $u = 10$, $N_1 = N_2 = 8$ equals the number of signs on each side of the median. Table M tells us that the critical value of u for $P = 0.05$ is 5, that is, there is a probability of 0.05 of obtaining as few as 5 runs or fewer. If we choose $\alpha = 0.05$ we should accept H_0. There is no evidence that the fluctuations are other than random ones.

15.6. The Use of Ranks

Often data, not available as measurements, are arranged in order of relative magnitude or importance, or according to some other criterion of rank. Flavors might be ordered according to satisfaction in taste by an expert taster, or young women might be ranked according to beauty by a board of judges. Also, it is often desirable to arrange measured items according to rank. Thus, the following scores in a history test could be assigned ranks as indicated:

Score	72	79	93	55	84	67	76	81	61	82
Rank	4	6	10	1	9	3	5	7	2	8

Here the lowest score, 55, is assigned the rank of 1; the next highest, 61, the rank of 2; and so on. When two or more variates are equal, it is customary to indicate the tie in rank by assigning to each the arithmetic mean of the rank which these items would have if they were in the same relative position with respect to the other items but not equal to one another. Thus, the following scores would be ranked as shown:

Score	70	76	57	70	76	83	84	76	94	62
Rank	$7\frac{1}{2}$	5	10	$7\frac{1}{2}$	5	3	2	5	1	9

In this illustration, the ranks 4, 5, and 6 are divided among the three variates 76, 76, and 76 by giving each a rank of $\frac{1}{3}(4 + 5 + 6)$, or 5. Similarly, the rank assigned to each of the two values 70 and 70 is $\frac{1}{2}(7 + 8)$, or $7\frac{1}{2}$.

The arithmetic mean of ranks. We recall from elementary algebra that the sum, S, of N terms of an arithmetic progression,

$$a + (a + d) + (a + 2d) + \cdots + [a + (N - 1)d],$$

where a is the first term and d the common difference, is given by the formula

$$S = \frac{N}{2}(a + l),$$

where l is the Nth or last term, $a + (N - 1)d$. It follows, then, that the sum of the first N integers 1, 2, 3, ..., N is merely $\frac{N}{2}(1 + N)$. The mean value, $\bar{R}$, of a set of nonduplicate ranks 1, 2, 3, ..., N is merely $\frac{1}{N}$th their sum, so that

$$\bar{R} = \frac{1}{N}\left[\frac{N}{2}(1 + N)\right]$$

$$= \frac{1}{2}(N + 1). \tag{15.3}$$

Obviously, the arithmetic mean of the integers from 1 to 19 is 10, and of the integers from 1 to 20 is $10\frac{1}{2}$. It should be observed that the presence of duplicate ranks will not affect the value of the mean as given by (15.3). Why?

The standard deviation of ranks. It is possible to establish, by mathematical induction, the formula

$$1^2 + 2^2 + 3^2 + \cdots + N^2 = \frac{N(N + 1)(2N + 1)}{6}. \tag{15.4}$$

By means of this and Formula (2.7), the variance of the ranks

$$s_R^2 = \frac{(N + 1)(2N + 1)}{6} - \frac{(N + 1)^2}{4}$$

$$= \frac{N^2 - 1}{12}. \tag{15.5}$$

Hence the standard deviation of N non-duplicate ranks is $\left(\dfrac{N^2 - 1}{12}\right)^{\frac{1}{2}}$.

The presence of tied ranks will introduce some error into Formula (15.5), but this will be slight if only a few duplicates are present, say, fewer than 20 per cent of all the ranks.

15.7. The Wilcoxon Rank-Sum Test

Suppose that we have two samples, $x_1, x_2, ..., x_{N_1}$, and $y_1, y_2, ..., y_{N_2}$, and arrange the combined samples of $N_1 + N_2$ observations in order of increasing size with appropriate ranks $1, 2, 3, ..., (N_1 + N_2)$. Let T' be the sum of the ranks in the smaller sample, or in either if $N_1 = N_2$. If N_1 is the number in the smaller sample

$$1 + 2 + 3 + \cdots + N_1 \leqq T' \leqq (N_2 + 1) + (N_2 + 2) + \cdots + (N_2 + N_1).$$

Since the sums on the left and the right are those of an arithmetic progression

$$\frac{N_1}{2}(N_1 + 1) \leqq T' \leqq \frac{N_1}{2}(2N_2 + N_1 + 1).$$

The distribution of T' under the hypothesis $H_0 : f_1(x) = f_2(x)$, has been derived and useful tables constructed. Table N is an abbreviated form of these tables.

Example. An examination covering a certain topic is given to pupils of two classes in order to compare the effectiveness of their teachers. The scores follow:

    Teacher A (x): 81 75 92 78 87 83 94 73 79 82
                    88 72 81 97 84 67 63 77 84 66

    Teacher B (y): 77 63 75 84 85 68 70 73 90 82 76 65

We arrange the combined sample of 32 scores according to rank.

63	63	65	66	67	68	70	72	73	73	75	75	76	77	77	78
$1\frac{1}{2}$	$1\frac{1}{2}$	3	4	5	6	7	8	$9\frac{1}{2}$	$9\frac{1}{2}$	$11\frac{1}{2}$	$11\frac{1}{2}$	13	$14\frac{1}{2}$	$14\frac{1}{2}$	16

79	81	81	82	82	83	84	84	84	85	87	88	90	92	94	97
17	$18\frac{1}{2}$	$18\frac{1}{2}$	$20\frac{1}{2}$	$20\frac{1}{2}$	22	24	24	24	26	27	28	29	30	31	32

The sum T', of the ranks of the smaller sample, those ranks corresponding to the 12 underlined scores, is 165.5. Since the sample sizes exceed 10 we approximate to the probability for as small a sum as 165.5 or smaller by means of the normal curve. Refer to the note at the bottom of Table N.

$$z = \frac{165.5 + 0.5 - \dfrac{12}{2}(33)}{\left(\dfrac{12 \times 20 \times 33}{12}\right)^{\frac{1}{2}}} = -1.24.$$

Then $P(T' < 165.5) = 0.1075$. Thus the sum of the ranks for Teacher B, although smaller than the average sum expected for equally effective teachers, is not small enough to refute the hypothesis of equal effectiveness at the five per cent level. Here we are taking as our alternative hypothesis $H_1:f_2(x) = f_1(x - x_0)$, that is, that the population $f_2(x)$ [pupils of Teacher B] is merely the population $f_1(x)$ [pupils of Teacher A] displaced to the left a distance x_0. This alternative calls for a one-tail (left-tail) test. A two-tail test, $H_1:f_2(x) = f_1(x \pm x_0)$ is possible.

15.8. Correlation of Ranked Data

The amount of association between two sets of paired ranks is measured by making use of the squared differences in rank. If d_i denotes the difference, $x_i - y_i$, between two paired ranks, we have

$$s_d^2 = \frac{1}{N} \sum (d_i - \bar{d})^2$$

$$= \frac{1}{N} \sum [(x_i - y_i) - (\bar{x} - \bar{y})]^2$$

$$= \frac{1}{N} \sum [(x_i - \bar{x}) - (y_i - \bar{y})]^2$$

$$= \frac{1}{N} \sum (x_i - \bar{x})^2 - \frac{2}{N} \sum (x_i - \bar{x})(y_i - \bar{y}) + \frac{1}{N} \sum (y_i - \bar{y})^2.$$

From (13.9) it follows that the middle term

$$\frac{2}{N} \sum (x - \bar{x})(y - \bar{y}) = 2rs_x s_y.$$

Substituting this result above, we get:

$$s_d^2 = s_x^2 - 2rs_x s_y + s_y^2. \tag{15.6}$$

If the x's and y's represent ranks, then both the ordered sets $x_1, x_2, x_3, ..., x_N$, and $y_1, y_2, y_3, ..., y_N$ are the same as the integers 1, 2, 3, ..., N. Thus

$$s_R^2 = s_x^2 = s_y^2,$$

and by (15.5)

$$s_R^2 = \frac{N^2 - 1}{12}.$$ (15.7)

It follows that (15.6) can be written

$$s_d^2 = 2s_R^2(1 - r),$$

whence

$$r = 1 - \frac{s_d^2}{2s_R^2},$$

and by (15.7)

$$r = 1 - \frac{6s_d^2}{N^2 - 1}.$$ (15.8)

Also

$$s_d^2 = \frac{1}{N} \sum (d_i - \bar{d})^2$$

$$= \frac{1}{N} \sum [(x_i - y_i) - (\bar{x} - \bar{y})]^2.$$

But for the two sets of identical ranks, $\bar{x} = \bar{y}$; hence,

$$s_d^2 = \frac{1}{N} \sum (x_i - y_i)^2 = \frac{1}{N} \sum d^2.$$

We may now write (15.8) as

$$r_R = 1 - \frac{6 \Sigma d^2}{N(N^2 - 1)}.$$ (15.9)

The symbol r_R, on the left of (15.9) is used to indicate the coefficient of correlation from ranks. The formula defines the *Spearman formula* for rank correlation.

Since r_R is merely a form of r for a special set of x's and y's, its value is restricted to the interval, -1 to $+1$. When the two sets of corresponding ranks are identical, each $d_i = 0$, the correlation is perfect, and $r_R = 1$. When high ranks are generally associated with low ranks, r_R is negative.

The use of Formula (15.9) is illustrated with the data of Table 13–1, ranked as shown in Table 15–1. The value of $r_R = -0.49$ for the ranks in that table.

Table O of the Supplementary Tables may be used to test the hypothesis of independence, that is, that the population rank correlation coefficient $\rho_R = 0$. The two-tail critical value for $P = 0.05$ and $N = 24$ is found from this table to be 0.41. Since this is exceeded numerically, by the value, 0.49, found above, we should have to reject the null hypothesis.

Although the Pearson r was used as a basis for developing the Spearman measure r_R, we make no assumption concerning the underlying distribution of x and y in applying this statistic to practical problems. It is a nonparametric measure. The use of ranks is advantageous when only ordinal data are available, or when the variates cannot be assumed to stem from a normal bivariate population. See, for example, Exercises 12 and 13.

Table 15–1

	Birth rate rank x_i	Property value rank y_i	$d_i = x_i - y_i$	d_i^2
Connecticut	1	20	−19	361
Massachusetts	2	14	−12	144
New York	3	16	−13	169
D. Columbia	4	21	−17	289
California	5	23	−18	324
N. Hampshire	6	12	−6	36
Vermont	7	5	2	4
Oregon	8	24	−16	256
Ohio	9	11	−2	4
Washington	10	19	−9	81
Maine	11	6	5	25
Pennsylvania	12	13	−1	1
Indiana	13	10	3	9
Wisconsin	14	8	6	36
Kansas	15	18	−3	9
Maryland	16	7	9	81
Michigan	17	9	8	64
Minnesota	18	17	1	1
Nebraska	19	22	−3	9
Kentucky	20	2	18	324
Virginia	21	4	17	289
S. Carolina	22	1	21	441
N. Carolina	23	3	20	400
Utah	24	15	9	81
				3438

$$r_R = 1 - \frac{6(3438)}{24(576 - 1)}$$
$$= -0.49.$$

EXERCISES

1. A variable x, has a nonnormal distribution with $\sigma = 5$. Use the Tchebycheff inequality to answer the following: (a) What can you say about the probability that x deviates from the mean by as much as (1) 15? (2) 10? (3) 5? (b) What would the answers be if you knew that the distribution was normal?

2. A variable x, has a nonnormal distribution with $\mu = 80$ and $\sigma = 10$. Use the Tchebycheff inequality to answer the following: What can you say about the probability that x lies between (a) 60 and 100? (b) 65 and 95?

3. Apply the sign test to Exercise 8.36 (Gains in weight of pigs).

4. Apply the sign test to Exercise 8.37 (Test scores of pupils under two teachers).

5. A comparison of magnesium (gm per mm) was made by two methods, the U.S.P. XIV and the E.D.T.A. Apply the sign test to see if there is a significant difference (5%) between the two.

Sample No:	1	2	3	4	5	6	7	8
U.S.P. XIV:	1.100	1.048	1.056	1.033	1.045	1.098	1.053	1.039
E.D.T.A.:	1.094	1.045	1.049	1.034	1.044	1.093	1.048	1.045

6. Two samples of 24 observations each were paired and the signs of the corresponding differences, d, were counted. There were 5 plus signs, 17 minus signs, and 2 zeros. Is there indication, say at the 10 per cent level, that one population is different from the other?

7. Apply the run test for randomness to the data of Exercise 8.16 (weekly average of number of parts inspected).

8. Gauge readings were made at equal intervals of time. The successive readings follow. Use the run test to see if they may be assumed to stem from the same population.

$$
\begin{array}{ccccccccc}
15.4 & 15.9 & 14.4 & 14.8 & 17.0 & 16.3 & 15.0 & 14.2 & 16.5 \\
15.3 & 16.2 & 15.1 & 14.8 & 17.2 & 16.5 & 15.7 & 14.3 & 14.0 \\
15.8 & 16.1 & 14.3 & 13.8 & 15.2 & 14.3 & 13.9 & 15.7 & 14.8
\end{array}
$$

9. Let X and Y represent two sets of observations as shown below. Apply the rank-sum test to see if they stem from the same population.

$$
\begin{array}{lccccccc}
X: & 20.2 & 24.3 & 21.8 & 25.9 & 30.2 & 26.3 & 22.0 \\
Y: & 26.2 & 29.4 & 35.7 & 39.8 & 41.2 & 25.3 &
\end{array}
$$

10. Suppose that in Exercise 8 the first 15 numbers represent gauge readings made under one set of conditions and that the remaining 12 numbers, readings under a second set of conditions. By means of the rank-sum, test the hypothesis that the two sets of conditions do not affect appreciably the magnitudes of the readings.

11. The numbers of minutes required for two groups of workers to assemble the parts of an instrument are as follows:

Group X: 16 21 14 15 17 25 18 19 17 31 19 16 28 15 17
Group Y: 16 16 18 15 13 19 20 18 18 14 13 18

Use the rank-sum test to see if the populations of times for these two types of workers differ (10% level).

12. In a baking contest 8 apple pies were ranked by two judges as follows:

$$\begin{array}{lcccccccc}
\text{Judge A:} & 8 & 2 & 4 & 3 & 7 & 6 & 5 & 1 \\
\text{Judge B:} & 5 & 1 & 3 & 7 & 8 & 6 & 2 & 4
\end{array}$$

What can you say about the extent of agreement between the two judges? On the basis of this contest could you conclude that these judges generally agree ($\alpha = 0.10$) in their ranking in similar contests?

13. In a mathematics examination the order in which the papers were passed in and their grades follow:

1	2	3	4	5	6	7	8	9	10	11	12	13	14	15	16	17
92	99	46	87	96	61	88	90	69	86	58	85	65	93	71	93	68

Find the Spearman rank correlation coefficient. Would you accept or reject the hypothesis $H_0 : \rho_R = 0$ with $\alpha = 0.05$?

14. The I.Q.'s and the grades in Statistics of 10 students were as follows:

$$\begin{array}{lcccccccccc}
\text{I.Q.:} & 108 & 139 & 90 & 100 & 120 & 80 & 103 & 95 & 125 & 110 \\
\text{Grade:} & 90 & 100 & 92 & 82 & 98 & 50 & 70 & 80 & 94 & 84
\end{array}$$

Compute r_R. Are I.Q.'s and grades in statistics independent of each other?

15. Assume that the scores of a class of students in a mathematics and in a physics test are as shown below. Compute r_R by means of Spearman's formula. Note the tied ranks, but neglect their influence on the value of r_R. Test $H_0 : \rho_R = 0$, with $\alpha = 0.05$.

$$\begin{array}{lcccccccc}
\text{Mathematics:} & 92 & 76 & 62 & 85 & 74 & 52 & 83 & 79 \\
\text{Physics:} & 90 & 80 & 68 & 71 & 64 & 63 & 95 & 83
\end{array}$$

$$\begin{array}{lccccccc}
\text{Mathematics:} & 75 & 84 & 68 & 51 & 73 & 77 & 66 \\
\text{Physics:} & 77 & 73 & 65 & 55 & 72 & 80 & 60
\end{array}$$

16. A group of students were ranked according to general school ability and according to personality. Compute r_R by means of Spearman's formula. Test $H_0 : \rho_R = 0$, with $\alpha = 0.10$.

$$\begin{array}{lcccccccccc}
\text{Ability:} & 1 & 2 & 3 & 4 & 5 & 6 & 7 & 8 & 9 & 10 \\
\text{Personality:} & 4 & 17 & 8 & 19 & 11 & 10 & 15 & 9 & 12 & 2
\end{array}$$

$$\begin{array}{lcccccccccc}
\text{Ability:} & 11 & 12 & 13 & 14 & 15 & 16 & 17 & 18 & 19 & 20 \\
\text{Personality:} & 1 & 7 & 5 & 14 & 20 & 18 & 13 & 6 & 3 & 16
\end{array}$$

REFERENCES

General

1. Croxton, F. E., and D. J. Cowden, *Applied General Statistics*, Second Edition, Prentice-Hall, Inc., Englewood Cliffs, N.J., 1955.

2. Dixon, W. J., and F. J. Massey, Jr., *Introduction to Statistical Analysis*, Second Edition, McGraw-Hill, New York, 1957.

3. Fisher, R. A., *Statistical Methods for Research Workers*, Twelfth Edition, Oliver and Boyd, Edinburgh, 1954.

4. Grant, E. L., *Statistical Quality Control*, Second Edition, McGraw-Hill, New York, 1952.

5. Hald, A., *Statistical Theory with Engineering Applications*, Wiley, New York, 1952.

6. Hill, A. B., *Principles of Medical Statistics*, Sixth Edition, Oxford, New York, 1955.

7. Kendall, M. G., *Rank Correlation Methods*, Charles Griffin, London, 1948.

8. Siegel, S., *Nonparametric Statistics for the Behavioral Sciences*, McGraw-Hill, New York, 1956.

9. Snedecor, G. W., *Statistical Methods*, Fifth Edition, Collegiate Press, Ames, Iowa, 1956.

10. Wallis, W. A., and H. V. Roberts, *Statistics—A New Approach*, The Free Press, Glencoe, Ill., 1956.

11. Wilks, S. S., *Elementary Statistical Analysis*, Princeton University Press, 1948.

12. Yule, G. U., and M. G. Kendall, *An Introduction to the Theory of Statistics*, Fourteenth Edition, Charles Griffin, London, 1950.

Mathematical Statistics

13. Cramer, H., *Mathematical Methods of Statistics*, Princeton University Press, 1946.

14. Feller, W., *An Introduction to Probability Theory and Its Applications*, Volume I, Second Edition, Wiley, New York, 1957.

15. Hoel, P. G., *Introduction to Mathematical Statistics*, Second Edition, Wiley, New York, 1954.

16. Kendall, M. G., *The Advanced Theory of Statistics* (Two volumes), Charles Griffin, London, 1947, 1948.

17. Kenney. J. F., and E. S. Keeping, *Mathematics of Statistics*, Van Nostrand, New York, Part I, Third Edition, 1954; Part 2, Second Edition, 1951.

18. Mood, A. M., *Introduction to the Theory of Statistics*, McGraw-Hill, New York, 1950.

Tables

19. David, F. N., *Tables of the Ordinates and Probability Integral of the Distribution of the Correlation Coefficient in Small Samples*, Cambridge University Press, 1938.

20. Fisher, R. A., and F. Yates, *Statistical Tables for Biological, Agricultural, and Medical Research*, Fifth Edition, Oliver and Boyd, Edinburgh, 1957.

21. Harvard University Computation Laboratory, *Tables of the Cumulative Binomial Probability Distribution*, Cambridge, 1955.

22. Mainland, D., L. Herrera, and M. I. Sutcliffe, *Statistical Tables for Use with Binomial Samples: Contingency Tests, Confidence Limits, and Samples Size Estimates*, Department of Medical Statistics, New York University College of Medicine, New York, 1956.

23. Molina, E. C., *Poisson's Exponential Binomial Limit*, Van Nostrand, New York, 1947.

24. National Bureau of Standards, *Tables of the Binomial Probability Distribution*, Washington, D.C., 1949.

25. Pearson, E. S., and Hartley, H. O., Editors, *Biometrika Tables for Statisticians*, (Two Volumes), Second Edition, Cambridge University Press, 1958.

26. Rand Corporation, *A Million Random Digits*, The Free Press, Glencoe, Ill., 1955.

Journals

27. *The Annals of Mathematical Statistics*, Official journal of the Institute of Mathematical Statistics, A. H. Bowker, Treasurer, Department of Statistics, Stanford University, Stanford, Calif.

28. *Biometrics*, published by the American Statistical Association for its Biometric Section, 1757 K St., N.W., Washington, D.C.

29. *Biometrika*, issued by the Biometrika Office, University College, London.

30. *Journal of the American Statistical Association*, 1757 K St., N.W., Washington, D.C.

31. *Psychometrika*, Official journal of the Psychometric Society, Colorado Springs, Colo.

MATHEMATICAL TABLES

CHIEFLY TO FOUR SIGNIFICANT DIGITS

Table A—SQUARES

TO FOUR SIGNIFICANT DIGITS
Square roots may be found by inverse interpolation.

N	.00	.01	.02	.03	.04	.05	.06	.07	.08	.09
1.0	1.000	1.020	1.040	1.061	1.082	1.102	1.124	1.145	1.166	1.188
1.1	1.210	1.232	1.254	1.277	1.300	1.322	1.346	1.369	1.392	1.416
1.2	1.440	1.464	1.488	1.513	1.538	1.562	1.588	1.613	1.638	1.664
1.3	1.690	1.716	1.742	1.769	1.796	1.822	1.850	1.877	1.904	1.932
1.4	1.960	1.988	2.016	2.045	2.074	2.102	2.132	2.161	2.190	2.220
1.5	2.250	2.280	2.310	2.341	2.372	2.402	2.434	2.465	2.496	2.528
1.6	2.560	2.592	2.624	2.657	2.690	2.722	2.756	2.789	2.822	2.856
1.7	2.890	2.924	2.958	2.993	3.028	3.062	3.098	3.133	3.168	3.204
1.8	3.240	3.276	3.312	3.349	3.386	3.422	3.460	3.497	3.534	3.572
1.9	3.610	3.648	3.686	3.725	3.764	3.802	3.842	3.881	3.920	3.960
2.0	4.000	4.040	4.080	4.121	4.162	4.202	4.244	4.285	4.326	4.368
2.1	4.410	4.452	4.494	4.537	4.580	4.622	4.666	4.709	4.752	4.796
2.2	4.840	4.884	4.928	4.973	5.018	5.062	5.108	5.153	5.198	5.244
2.3	5.290	5.336	5.382	5.429	5.476	5.522	5.570	5.617	5.664	5.712
2.4	5.760	5.808	5.856	5.905	5.954	6.002	6.052	6.101	6.150	6.200
2.5	6.250	6.300	6.350	6.401	6.452	6.502	6.554	6.605	6.656	6.708
2.6	6.760	6.812	6.864	6.917	6.970	7.022	7.076	7.129	7.182	7.236
2.7	7.290	7.344	7.398	7.453	7.508	7.562	7.618	7.673	7.728	7.784
2.8	7.840	7.896	7.952	8.009	8.066	8.122	8.180	8.237	8.294	8.352
2.9	8.410	8.468	8.526	8.585	8.644	8.702	8.762	8.821	8.880	8.940
3.0	9.000	9.060	9.120	9.181	9.242	9.302	9.364	9.425	9.486	9.548
3.1	9.610	9.672	9.734	9.797	9.860	9.922	9.986	10.05	10.11	10.18
3.2	10.24	10.30	10.37	10.43	10.50	10.56	10.63	10.69	10.76	10.82
3.3	10.89	10.96	11.02	11.09	11.16	11.22	11.29	11.36	11.42	11.49
3.4	11.56	11.63	11.70	11.76	11.83	11.90	11.97	12.04	12.11	12.18
3.5	12.25	12.32	12.39	12.46	12.53	12.60	12.67	12.74	12.82	12.89
3.6	12.96	13.03	13.10	13.18	13.25	13.32	13.40	13.47	13.54	13.62
3.7	13.69	13.76	13.84	13.91	13.99	14.06	14.14	14.21	14.29	14.36
3.8	14.44	14.52	14.59	14.67	14.75	14.82	14.90	14.98	15.05	15.13
3.9	15.21	15.29	15.37	15.44	15.52	15.60	15.68	15.76	15.84	15.92
4.0	16.00	16.08	16.16	16.24	16.32	16.40	16.48	16.56	16.65	16.73
4.1	16.81	16.89	16.97	17.06	17.14	17.22	17.31	17.39	17.47	17.56
4.2	17.64	17.72	17.81	17.89	17.98	18.06	18.15	18.23	18.32	18.40
4.3	18.49	18.58	18.66	18.75	18.84	18.92	19.01	19.10	19.18	19.27
4.4	19.36	19.45	19.54	19.62	19.71	19.80	19.89	19.98	20.07	20.16
4.5	20.25	20.34	20.43	20.52	20.61	20.70	20.79	20.88	20.98	21.07
4.6	21.16	21.25	21.34	21.44	21.53	21.62	21.72	21.81	21.90	22.00
4.7	22.09	22.18	22.28	22.37	22.47	22.56	22.66	22.75	22.85	22.94
4.8	23.04	23.14	23.23	23.33	23.43	23.52	23.62	23.72	23.81	23.91
4.9	24.01	24.11	24.21	24.30	24.40	24.50	24.60	24.70	24.80	24.90
5.0	25.00	25.10	25.20	25.30	25.40	25.50	25.60	25.70	25.81	25.91
N	.00	.01	.02	.03	.04	.05	.06	.07	.08	.09

Table A (*Continued*)

N	.00	.01	.02	.03	.04	.05	.06	.07	.08	.09
5.0	25.00	25.10	25.20	25.30	25.40	25.50	25.60	25.70	25.81	25.91
5.1	26.01	26.11	26.21	26.32	26.42	26.52	26.63	26.73	26.83	26.94
5.2	27.04	27.14	27.25	27.35	27.46	27.56	27.67	27.77	27.88	27.98
5.3	28.09	28.20	28.30	28.41	28.52	28.62	28.73	28.84	28.94	29.05
5.4	29.16	29.27	29.38	29.48	29.59	29.70	29.81	29.92	30.03	30.14
5.5	30.25	30.36	30.47	30.58	30.69	30.80	30.91	31.02	31.14	31.25
5.6	31.36	31.47	31.58	31.70	31.81	31.92	32.04	32.15	32.26	32.38
5.7	32.49	32.60	32.72	32.83	32.95	33.06	33.18	33.29	33.41	33.52
5.8	33.64	33.76	33.87	33.99	34.11	34.22	34.34	34.46	34.57	34.69
5.9	34.81	34.93	35.05	35.16	35.28	35.40	35.52	35.64	35.76	35.88
6.0	36.00	36.12	36.24	36.36	36.48	36.60	36.72	36.84	36.97	37.09
6.1	37.21	37.33	37.45	37.58	37.70	37.82	37.95	38.07	38.19	38.32
6.2	38.44	38.56	38.69	38.81	38.94	39.06	39.19	39.31	39.44	39.56
6.3	39.69	39.82	39.94	40.07	40.20	40.32	40.45	40.58	40.70	40.83
6.4	40.96	41.09	41.22	41.34	41.47	41.60	41.73	41.86	41.99	42.12
6.5	42.25	42.38	42.51	42.64	42.77	42.90	43.03	43.16	43.30	43.43
6.6	43.56	43.69	43.82	43.96	44.09	44.22	44.36	44.49	44.62	44.76
6.7	44.89	45.02	45.16	45.29	45.43	45 56	45.70	45.83	45.97	46.10
6.8	46.24	46.38	46.51	46.65	46.79	46.92	47.06	47.20	47.33	47.47
6.9	47.61	47.75	47.89	48.02	48.16	48.30	48.44	48 58	49.72	40.00
7.0	49.00	49.14	49.28	49.42	49.56	49.70	49.84	49.98	50.13	50.27
7.1	50.41	50.55	50.69	50.84	50.98	51.12	51.27	51.41	51.55	51.70
7.2	51.84	51.98	52.13	52.27	52.42	52.56	52.71	52.85	53.00	53.14
7.3	53.29	53.44	53.58	53.73	53.88	54.02	54.17	54.32	54.46	54.61
7.4	54.76	54.91	55.06	55.20	55.35	55.50	55.65	55.80	55.95	56.10
7.5	56.25	56.40	56.55	56.70	56.85	57.00	57.15	57.30	57.46	57.61
7.6	57.76	57.91	58.06	58.22	58.37	58.52	58.68	58.83	58.98	59.14
7.7	59.29	59.44	59.60	59.75	59.91	60.06	60.22	60.37	60.53	60.68
7.8	60.84	61.00	61.15	61.31	61.47	61.62	61.78	61.94	62.09	62.25
7.9	62.41	62.57	62.73	62.88	63.04	63.20	63.36	63.52	63.68	63.84
8.0	64.00	64.16	64.32	64.48	64.64	64.80	64.96	65.12	65.29	65.45
8.1	65.61	65.77	65.93	66.10	66.26	66.42	66.59	66.75	66.91	67.08
8.2	67.24	67.40	67.57	67.73	67.90	68.06	68.23	68.39	68.56	68.72
8.3	68.89	69.06	69.22	69.39	69.56	69.72	69.89	70.06	70.22	70.39
8.4	70.56	70.73	70.90	71.06	71.23	71.40	71.57	71.74	71.91	72.08
8.5	72.25	72.42	72.59	72.76	72.93	73.10	73.27	73.44	73.62	73.79
8.6	73.96	74.13	74.30	74.48	74.65	74.82	75.00	75.17	75.34	75.52
8.7	75.69	75.86	76.04	76.21	76.39	76.56	76.74	76.91	77.09	77.26
8.8	77.44	77.62	77.79	77.97	78.15	78.32	78.50	78.68	78.85	79.03
8.9	79.21	79.39	79.57	79.74	79.92	80.10	80.28	80.46	80.64	80.82
9.0	81.00	81.18	81.36	81.54	81.72	81.90	82.08	82.26	82.45	82.63
9.1	82.81	82.99	83.17	83.36	83.54	83.72	83.91	84.09	84.27	84.46
9.2	84.64	84.82	85.01	85.19	85.38	85.56	85.75	85.93	86.12	86.30
9.3	86.49	86.68	86.86	87.05	87.24	87.42	87.61	87.80	87.98	88.17
9.4	88.36	88.55	88.74	88.92	89.11	89.30	89.49	89.68	89.87	90.06
9.5	90.25	90.44	90.63	90.82	91.01	91.20	91.39	91.58	91.78	91.97
9.6	92.16	92.35	92.54	92.74	92.93	93.12	93.32	93.51	93.70	93.90
9.7	94.09	94.28	94.48	94.67	94.87	95.06	95.26	95.45	95.65	95.84
9.8	96.04	96.24	96.43	96.63	96.83	97.02	97.22	97.42	97.61	97.81
9.9	98.01	98.21	98.41	98.60	98.80	99.00	99.20	99.40	99.60	99.80
N	.00	.01	.02	.03	.04	.05	.06	.07	.08	.09

MATHEMATICAL TABLES

Table B—Reciprocals
To Four Significant Digits

N	.00	.01	.02	.03	.04	.05	.06	.07	.08	.09
1.0	1.0000	.9901	.9804	.9709	.9615	.9524	.9434	.9346	.9259	.9174
1.1	.9091	.9009	.8929	.8850	.8772	.8696	.8621	.8547	.8475	.8403
1.2	.8333	.8264	.8197	.8130	.8065	.8000	.7937	.7874	.7812	.7752
1.3	.7692	.7634	.7576	.7519	.7463	.7407	.7353	.7299	.7246	.7194
1.4	.7143	.7092	.7042	.6993	.6944	.6897	.6849	.6803	.6757	.6711
1.5	.6667	.6623	.6579	.6536	.6494	.6452	.6410	.6369	.6329	.6289
1.6	.6250	.6211	.6173	.6135	.6098	.6061	.6024	.5988	.5952	.5917
1.7	.5882	.5848	.5814	.5780	.5747	.5714	.5682	.5650	.5618	.5587
1.8	.5556	.5525	.5495	.5464	.5435	.5405	.5376	.5348	.5319	.5291
1.9	.5263	.5236	.5208	.5181	.5155	.5128	.5102	.5076	.5051	.5025
2.0	.5000	.4975	.4950	.4926	.4902	.4878	.4854	.4831	.4808	.4785
2.1	.4762	.4739	.4717	.4695	.4673	.4651	.4630	.4608	.4587	.4566
2.2	.4545	.4525	.4505	.4484	.4464	.4444	.4425	.4405	.4386	.4367
2.3	.4348	.4329	.4310	.4292	.4274	.4255	.4237	.4219	.4202	.4184
2.4	.4167	.4149	.4132	.4115	.4098	.4082	.4065	.4049	.4032	.4016
2.5	.4000	.3984	.3968	.3953	.3937	.3922	.3906	.3891	.3876	.3861
2.6	.3846	.3831	.3817	.3802	.3788	.3774	.3759	.3745	.3731	.3717
2.7	.3704	.3690	.3676	.3663	.3650	.3636	.3623	.3610	.3597	.3584
2.8	.3571	.3559	.3546	.3534	.3521	.3509	.3497	.3484	.3472	.3460
2.9	.3448	.3436	.3425	.3413	.3401	.3390	.3378	.3367	.3356	.3344
3.0	.3333	.3322	.3311	.3300	.3289	.3279	.3268	.3257	.3247	.3236
3.1	.3226	.3215	.3205	.3195	.3185	.3175	.3165	.3155	.3145	.3135
3.2	.3125	.3115	.3106	.3096	.3086	.3077	.3067	.3058	.3049	.3040
3.3	.3030	.3021	.3012	.3003	.2994	.2985	.2976	.2967	.2959	.2950
3.4	.2941	.2933	.2924	.2915	.2907	.2899	.2890	.2882	.2874	.2865
3.5	.2857	.2849	.2841	.2833	.2825	.2817	.2809	.2801	.2793	.2786
3.6	.2778	.2770	.2762	.2755	.2747	.2740	.2732	.2725	.2717	.2710
3.7	.2703	.2695	.2688	.2681	.2674	.2667	.2660	.2653	.2646	.2639
3.8	.2632	.2625	.2618	.2611	.2604	.2597	.2591	.2584	.2577	.2571
3.9	.2564	.2558	.2551	.2545	.2538	.2532	.2525	.2519	.2513	.2506
4.0	.2500	.2494	.2488	.2481	.2475	.2469	.2463	.2457	.2451	.2445
4.1	.2439	.2433	.2427	.2421	.2415	.2410	.2404	.2398	.2392	.2387
4.2	.2381	.2375	.2370	.2364	.2358	.2353	.2347	.2342	.2336	.2331
4.3	.2326	.2320	.2315	.2309	.2304	.2299	.2294	.2288	.2283	.2278
4.4	.2273	.2268	.2262	.2257	.2252	.2247	.2242	.2237	.2232	.2227
4.5	.2222	.2217	.2212	.2208	.2203	.2198	.2193	.2188	.2183	.2179
4.6	.2174	.2169	.2165	.2160	.2155	.2151	.2146	.2141	.2137	.2132
4.7	.2128	.2123	.2119	.2114	.2110	.2105	.2101	.2096	.2092	.2088
4.8	.2083	.2079	.2075	.2070	.2066	.2062	.2058	.2053	.2049	.2045
4.9	.2041	.2037	.2033	.2028	.2024	.2020	.2016	.2012	.2008	.2004
5.0	.2000	.1996	.1992	.1988	.1984	.1980	.1976	.1972	.1969	.1965
N	.00	.01	.02	.03	.04	.05	.06	.07	.08	.09

Table B (*Continued*)

N	.00	.01	.02	.03	.04	.05	.06	.07	.08	.09
5.0	.2000	.1996	.1992	.1988	.1984	.1980	.1976	.1972	.1969	.1965
5.1	.1961	.1957	.1953	.1949	.1946	.1942	.1938	.1934	.1931	.1927
5.2	.1923	.1919	.1916	.1912	.1908	.1905	.1901	.1898	.1894	.1890
5.3	.1887	.1883	.1880	.1876	.1873	.1869	.1866	.1862	.1859	.1855
5.4	.1852	.1848	.1845	.1842	.1838	.1835	.1832	.1828	.1825	.1821
5.5	.1818	.1815	.1812	.1808	.1805	.1802	.1799	.1795	.1792	.1789
5.6	.1786	.1783	.1779	.1776	.1773	.1770	.1767	.1764	.1761	.1757
5.7	.1754	.1751	.1748	.1745	.1742	.1739	.1736	.1733	.1730	.1727
5.8	.1724	.1721	.1718	.1715	.1712	.1709	.1706	.1704	.1701	.1698
5.9	.1695	.1692	.1689	.1686	.1684	.1681	.1678	.1675	.1672	.1669
6.0	.1667	.1664	.1661	.1658	.1656	.1653	.1650	.1647	.1645	.1642
6.1	.1639	.1637	.1634	.1631	.1629	.1626	.1623	.1621	.1618	.1616
6.2	.1613	.1610	.1608	.1605	.1603	.1600	.1597	.1595	.1592	.1590
6.3	.1587	.1585	.1582	.1580	.1577	.1575	.1572	.1570	.1567	.1565
6.4	.1562	.1560	.1558	.1555	.1553	.1550	.1548	.1546	.1543	.1541
6.5	.1538	.1536	.1534	.1531	.1529	.1527	.1524	.1522	.1520	.1517
6.6	.1515	.1513	.1511	.1508	.1506	.1504	.1502	.1499	.1497	.1495
6.7	.1493	.1490	.1488	.1486	.1484	.1481	.1479	.1477	.1475	.1473
6.8	.1471	.1468	.1466	.1464	.1462	.1460	.1458	.1456	.1453	.1451
6.9	.1449	.1447	.1445	.1443	.1441	.1439	.1437	.1435	.1433	.1431
7.0	.1429	.1427	.1425	.1422	.1420	.1418	.1416	.1414	.1412	.1410
7.1	.1408	.1406	.1404	.1403	.1401	.1399	.1397	.1395	.1393	.1391
7.2	.1389	.1387	.1385	.1383	.1381	.1379	.1377	.1376	.1374	.1372
7.3	.1370	.1368	.1366	.1364	.1362	.1361	.1359	.1357	.1355	.1353
7.4	.1351	.1350	.1348	.1346	.1344	.1342	.1340	.1339	.1337	.1335
7.5	.1333	.1332	.1330	.1328	.1326	.1325	.1323	.1321	.1319	.1318
7.6	.1316	.1314	.1312	.1311	.1309	.1307	.1305	.1304	.1302	.1300
7.7	.1299	.1297	.1295	.1294	.1292	.1290	.1289	.1287	.1285	.1284
7.8	.1282	.1280	.1279	.1277	.1276	.1274	.1272	.1271	.1269	.1267
7.9	.1266	.1264	.1263	.1261	.1259	.1258	.1256	.1255	.1253	.1252
8.0	.1250	.1248	.1247	.1245	.1244	.1242	.1241	.1239	.1238	.1236
8.1	.1235	.1233	.1232	.1230	.1229	.1227	.1225	.1224	.1222	.1221
8.2	.1220	.1218	.1217	.1215	.1214	.1212	.1211	.1209	.1208	.1206
8.3	.1205	.1203	.1202	.1200	.1199	.1198	.1196	.1195	.1193	.1192
8.4	.1190	.1189	.1188	.1186	.1185	.1183	.1182	.1181	.1179	.1178
8.5	.1176	.1175	.1174	.1172	.1171	.1170	.1168	.1167	.1166	.1164
8.6	.1163	.1161	.1160	.1159	.1157	.1156	.1155	.1153	.1152	.1151
8.7	.1149	.1148	.1147	.1145	.1144	.1143	.1142	.1140	.1139	.1138
8.8	.1136	.1135	.1134	.1133	.1131	.1130	.1129	.1127	.1126	.1125
8.9	.1124	.1122	.1121	.1120	.1119	.1117	.1116	.1115	.1114	.1112
9.0	.1111	.1110	.1109	.1107	.1106	.1105	.1104	.1103	.1101	.1100
9.1	.1099	.1098	.1096	.1095	.1094	.1093	.1092	.1091	.1089	.1088
9.2	.1087	.1086	.1085	.1083	.1082	.1081	.1080	.1079	.1078	.1076
9.3	.1075	.1074	.1073	.1072	.1071	.1070	.1068	.1067	.1066	.1065
9.4	.1064	.1063	.1062	.1060	.1059	.1058	.1057	.1056	.1055	.1054
9.5	.1053	.1052	.1050	.1049	.1048	.1047	.1046	.1045	.1044	.1043
9.6	.1042	.1041	.1040	.1038	.1037	.1036	.1035	.1034	.1033	.1032
9.7	.1031	.1030	.1029	.1028	.1027	.1026	.1025	.1024	.1022	.1021
9.8	.1020	.1019	.1018	.1017	.1016	.1015	.1014	.1013	.1012	.1011
9.9	.1010	.1009	.1008	.1007	.1006	.1005	.1004	.1003	.1002	.1001
N	.00	.01	.02	.03	.04	.05	.06	.07	.08	.09

Table C—Ordinates of the Normal Curve

$$\phi(z) = \frac{1}{\sqrt{2\pi}}\, e^{-\frac{1}{2}z^2}$$

To Four Decimal Places

z	.00	.01	.02	.03	.04	.05	.06	.07	.08	.09
.0	.3989	.3989	.3989	.3988	.3986	.3984	.3982	.3980	.3977	.3973
.1	.3970	.3965	.3961	.3956	.3951	.3945	.3939	.3932	.3925	.3918
.2	.3910	.3902	.3894	.3885	.3876	.3867	.3857	.3847	.3836	.3825
.3	.3814	.3802	.3790	.3778	.3765	.3752	.3739	.3725	.3712	.3697
.4	.3683	.3668	.3653	.3637	.3621	.3605	.3589	.3572	.3555	.3538
.5	.3521	.3503	.3485	.3467	.3448	.3429	.3410	.3391	.3372	.3352
.6	.3332	.3312	.3292	.3271	.3251	.3230	.3209	.3187	.3166	.3144
.7	.3123	.3101	.3079	.3056	.3034	.3011	.2989	.2966	.2943	.2920
.8	.2897	.2874	.2850	.2827	.2803	.2780	.2756	.2732	.2709	.2685
.9	.2661	.2637	.2613	.2589	.2565	.2541	.2516	.2492	.2468	.2444
1.0	.2420	.2396	.2371	.2347	.2323	.2299	.2275	.2251	.2227	.2203
1.1	.2179	.2155	.2131	.2107	.2083	.2059	.2036	.2012	.1989	.1965
1.2	.1942	.1919	.1895	.1872	.1849	.1826	.1804	.1781	.1758	.1736
1.3	.1714	.1691	.1669	.1647	.1626	.1604	.1582	.1561	.1539	.1518
1.4	.1497	.1476	.1456	.1435	.1415	.1394	.1374	.1354	.1334	.1315
1.5	.1295	.1276	.1257	.1238	.1219	.1200	.1182	.1163	.1145	.1127
1.6	.1109	.1092	.1074	.1057	.1040	.1023	.1006	.0989	.0973	.0957
1.7	.0940	.0925	.0909	.0893	.0878	.0863	.0848	.0833	.0818	.0804
1.8	.0790	.0775	.0761	.0748	.0734	.0721	.0707	.0694	.0681	.0669
1.9	.0656	.0644	.0632	.0620	.0608	.0596	.0584	.0573	.0562	.0551
2.0	.0540	.0529	.0519	.0508	.0498	.0488	.0478	.0468	.0459	.0449
2.1	.0440	.0431	.0422	.0413	.0404	.0396	.0387	.0379	.0371	.0363
2.2	.0355	.0347	.0339	.0332	.0325	.0317	.0310	.0303	.0297	.0290
2.3	.0283	.0277	.0270	.0264	.0258	.0252	.0246	.0241	.0235	.0229
2.4	.0224	.0219	.0213	.0208	.0203	.0198	.0194	.0189	.0184	.0180
2.5	.0175	.0171	.0167	.0163	.0158	.0154	.0151	.0147	.0143	.0139
2.6	.0136	.0132	.0129	.0126	.0122	.0119	.0116	.0113	.0110	.0107
2.7	.0104	.0101	.0099	.0096	.0093	.0091	.0088	.0086	.0084	.0081
2.8	.0079	.0077	.0075	.0073	.0071	.0069	.0067	.0065	.0063	.0061
2.9	.0060	.0058	.0056	.0055	.0053	.0051	.0050	.0048	.0047	.0046
3.0	.0044	.0043	.0042	.0040	.0039	.0038	.0037	.0036	.0035	.0034
3.1	.0033	.0032	.0031	.0030	.0029	.0028	.0027	.0026	.0025	.0025
3.2	.0024	.0023	.0022	.0022	.0021	.0020	.0020	.0019	.0018	.0018
3.3	.0017	.0017	.0016	.0016	.0015	.0015	.0014	.0014	.0013	.0013
3.4	.0012	.0012	.0012	.0011	.0011	.0010	.0010	.0010	.0009	.0009
3.5	.0009	.0008	.0008	.0008	.0008	.0007	.0007	.0007	.0007	.0006
3.6	.0006	.0006	.0006	.0005	.0005	.0005	.0005	.0005	.0005	.0004
3.7	.0004	.0004	.0004	.0004	.0004	.0004	.0003	.0003	.0003	.0003
3.8	.0003	.0003	.0003	.0003	.0003	.0002	.0002	.0002	.0002	.0002
3.9	.0002	.0002	.0002	.0002	.0002	.0002	.0002	.0002	.0001	.0001
4.0	.0001	.0001	.0001	.0001	.0001	.0001	.0001	.0001	.0001	.0001
z	.00	.01	.02	.03	.04	.05	.06	.07	.08	.09

Table D—AREA UNDER THE NORMAL CURVE

$$A = \int_z^\infty \phi(z)\, dz$$

TO FOUR DECIMAL PLACES

z	.00	.01	.02	.03	.04	.05	.06	.07	.08	.09
.0	.5000	.4960	.4920	.4880	.4840	.4801	.4761	.4721	.4681	.4641
.1	.4602	.4562	.4522	.4483	.4443	.4404	.4364	.4325	.4286	.4247
.2	.4207	.4168	.4129	.4090	.4052	.4013	.3974	.3936	.3897	.3859
.3	.3821	.3783	.3745	.3707	.3669	.3632	.3594	.3557	.3520	.3483
.4	.3446	.3409	.3372	.3336	.3300	.3264	.3228	.3192	.3156	.3121
.5	.3085	.3050	.3015	.2981	.2946	.2912	.2877	.2843	.2810	.2776
.6	.2743	.2709	.2676	.2643	.2611	.2578	.2546	.2514	.2483	.2451
.7	.2420	.2389	.2358	.2327	.2296	.2266	.2236	.2206	.2177	.2148
.8	.2119	.2090	.2061	.2033	.2005	.1977	.1949	.1922	.1894	.1867
.9	.1841	.1814	.1788	.1762	.1736	.1711	.1685	.1660	.1635	.1611
1.0	.1587	.1562	.1539	.1515	.1492	.1469	.1446	.1423	.1401	.1379
1.1	.1357	.1335	.1314	.1292	.1271	.1251	.1230	.1210	.1190	.1170
1.2	.1151	.1131	.1112	.1093	.1075	.1056	.1038	.1020	.1003	.0985
1.3	.0968	.0951	.0934	.0918	.0901	.0885	.0869	.0853	.0838	.0823
1.4	.0808	.0793	.0778	.0764	.0749	.0735	.0721	.0708	.0694	.0681
1.5	.0668	.0655	.0643	.0630	.0618	.0606	.0594	.0582	.0571	.0559
1.6	.0548	.0537	.0526	.0516	.0505	.0495	.0485	.0475	.0465	.0455
1.7	.0446	.0436	.0427	.0418	.0409	.0401	.0392	.0384	.0375	.0367
1.8	.0359	.0351	.0344	.0336	.0329	.0322	.0314	.0307	.0301	.0294
1.9	.0287	.0281	.0274	.0268	.0262	.0256	.0250	.0244	.0239	.0233
2.0	.0228	.0222	.0217	.0212	.0207	.0202	.0197	.0192	.0188	.0183
2.1	.0179	.0174	.0170	.0166	.0162	.0158	.0154	.0150	.0146	.0143
2.2	.0139	.0136	.0132	.0129	.0125	.0122	.0119	.0116	.0113	.0110
2.3	.0107	.0104	.0102	.0099	.0096	.0094	.0091	.0089	.0087	.0084
2.4	.0082	.0080	.0078	.0075	.0073	.0071	.0069	.0068	.0066	.0064
2.5	.0062	.0060	.0059	.0057	.0055	.0054	.0052	.0051	.0049	.0048
2.6	.0047	.0045	.0044	.0043	.0041	.0040	.0039	.0038	.0037	.0036
2.7	.0035	.0034	.0033	.0032	.0031	.0030	.0029	.0028	.0027	.0026
2.8	.0026	.0025	.0024	.0023	.0023	.0022	.0021	.0021	.0020	.0019
2.9	.0019	.0018	.0018	.0017	.0016	.0016	.0015	.0015	.0014	.0014
3.0	.0013	.0013	.0013	.0012	.0012	.0011	.0011	.0011	.0010	.0010
3.1	.0010	.0009	.0009	.0009	.0008	.0008	.0008	.0008	.0007	.0007
3.2	.0007	.0007	.0006	.0006	.0006	.0006	.0006	.0005	.0005	.0005
3.3	.0005	.0005	.0005	.0004	.0004	.0004	.0004	.0004	.0004	.0003
3.4	.0003	.0003	.0003	.0003	.0003	.0003	.0003	.0003	.0003	.0002
3.5	.0002	.0002	.0002	.0002	.0002	.0002	.0002	.0002	.0002	.0002
3.6	.0002	.0002	.0001	.0001	.0001	.0001	.0001	.0001	.0001	.0001
3.7	.0001	.0001	.0001	.0001	.0001	.0001	.0001	.0001	.0001	.0001
3.8	.0001	.0001	.0001	.0001	.0001	.0001	.0001	.0001	.0001	.0001
3.9	.0000	.0000	.0000	.0000	.0000	.0000	.0000	.0000	.0000	.0000
z	.00	.01	.02	.03	.04	.05	.06	.07	.08	.09

Table E—VALUES OF FISHER'S *t*

TO THREE DECIMAL PLACES

P equals the probability of exceeding a given positive value of *t* corresponding to *n* degrees of freedom. (One-tailed probability.)

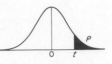

n \ P	.45	.40	.35	.20	.15	.10	.05	.025	.01	.005
1	.158	.325	.510	1.376	1.963	3.078	6.314	12.71	31.82	63.66
2	.142	.289	.445	1.061	1.386	1.886	2.920	4.303	6.965	9.925
3	.137	.277	.424	.978	1.250	1.638	2.353	3.182	4.541	5.841
4	.134	.271	.414	.941	1.190	1.533	2.132	2.776	3.747	4.604
5	.132	.267	.408	.920	1.156	1.476	2.015	2.571	3.365	4.032
6	.131	.265	.404	.906	1.134	1.440	1.943	2.447	3.143	3.707
7	.130	.263	.402	.896	1.119	1.415	1.895	2.365	2.998	3.499
8	.130	.262	.399	.889	1.108	1.397	1.860	2.306	2.896	3.355
9	.129	.261	.398	.883	1.100	1.383	1.833	2.262	2.821	3.250
10	.129	.260	.397	.879	1.093	1.372	1.812	2.228	2.764	3.169
11	.129	.260	.396	.876	1.088	1.363	1.796	2.201	2.718	3.106
12	.128	.259	.395	.873	1.083	1.356	1.782	2.179	2.681	3.055
13	.128	.259	.394	.870	1.079	1.350	1.771	2.160	2.650	3.012
14	.128	.258	.393	.868	1.076	1.345	1.761	2.145	2.624	2.977
15	.128	.258	.393	.866	1.074	1.341	1.753	2.131	2.602	2.947
16	.128	.258	.392	.865	1.071	1.337	1.746	2.120	2.583	2.921
17	.128	.257	.392	.863	1.069	1.333	1.740	2.110	2.567	2.898
18	.127	.257	.392	.862	1.067	1.330	1.734	2.101	2.552	2.878
19	.127	.257	.391	.861	1.066	1.328	1.729	2.093	2.539	2.861
20	.127	.257	.391	.860	1.064	1.325	1.725	2.086	2.528	2.845
21	.127	.257	.391	.859	1.063	1.323	1.721	2.080	2.518	2.831
22	.127	.256	.390	.858	1.061	1.321	1.717	2.074	2.508	2.819
23	.127	.256	.390	.858	1.060	1.319	1.714	2.069	2.500	2.807
24	.127	.256	.390	.857	1.059	1.318	1.711	2.064	2.492	2.797
25	.127	.256	.390	.856	1.058	1.316	1.708	2.060	2.485	2.787
26	.127	.256	.390	.856	1.058	1.315	1.706	2.056	2.479	2.779
27	.127	.256	.389	.855	1.057	1.314	1.703	2.052	2.473	2.771
28	.127	.256	.389	.855	1.056	1.313	1.701	2.048	2.467	2.763
29	.127	.256	.389	.854	1.055	1.311	1.699	2.045	2.462	2.756
30	.127	.256	.389	.854	1.055	1.310	1.697	2.042	2.457	2.750
40	.126	.255	.388	.851	1.050	1.303	1.684	2.021	2.423	2.704
60	.126	.254	.387	.848	1.046	1.296	1.671	2.000	2.390	2.660
120	.126	.254	.386	.845	1.041	1.289	1.658	1.980	2.358	2.617
∞	.126	.253	.385	.842	1.036	1.282	1.645	1.960	2.326	2.576
n \ P	.45	.40	.35	.20	.15	.10	.05	.025	.01	.005

Table E is abridged from Table IV of Fisher: *Statistical Methods for Research Workers*, Oliver and Boyd, Ltd., Edinburgh, by permission of the author and publishers.

Chiefly to Four Significant Digits

P equals the probability of exceeding a given value of χ^2 corresponding to n degrees of freedom.

P n	.99	.98	.95	.90	.50	.10	.05	.02	.01
1	.0002	.0006	.0039	.0158	.455	2.706	3.841	5.412	6.635
2	.0201	.0404	.103	.211	1.386	4.605	5.991	7.824	9.210
3	.115	.185	.352	.584	2.366	6.251	7.815	9.837	11.34
4	.297	.429	.711	1.064	3.357	7.779	9.488	11.67	13.28
5	.554	.752	1.145	1.610	4.351	9.236	11.07	13.39	15.09
6	.872	1.134	1.635	2.204	5.348	10.64	12.59	15.03	16.81
7	1.239	1.564	2.167	2.833	6.346	12.02	14.07	16.62	18.48
8	1.646	2.032	2.733	3.490	7.344	13.36	15.51	18.17	20.09
9	2.088	2.532	3.325	4.168	8.343	14.68	16.92	19.68	21.67
10	2.558	3.059	3.940	4.865	9.342	15.99	18.31	21.16	23.21
11	3.053	3.609	4.575	5.578	10.34	17.28	19.68	22.62	24.72
12	3.571	4.178	5.226	6.301	11.34	18.55	21.03	24.05	26.22
13	4.107	4.765	5.892	7.042	12.34	19.81	22.36	25.47	27.69
14	4.660	5.368	6.571	7.790	13.34	21.06	23.68	26.87	29.14
15	5.229	5.985	7.261	8.547	14.34	22.31	25.00	28.26	30.58
16	5.812	6.614	7.962	9.312	15.34	23.54	26.30	29.63	32.00
17	6.408	7.255	8.672	10.08	16.34	24.77	27.59	31.00	33.41
18	7.015	7.906	9.390	10.86	17.34	25.99	28.87	32.35	34.80
19	7.633	8.567	10.12	11.65	18.34	27.20	30.14	33.69	36.19
20	8.260	9.237	10.85	12.44	19.34	28.41	31.41	35.02	37.57
21	8.897	9.915	11.59	13.24	20.34	29.62	32.67	36.34	38.93
22	9.542	10.60	12.34	14.04	21.34	30.81	33.92	37.66	40.29
23	10.20	11.29	13.09	14.85	22.34	32.01	35.17	38.97	41.64
24	10.86	11.99	13.85	15.66	23.34	33.20	36.42	40.27	42.98
25	11.52	12.70	14.61	16.47	24.34	34.38	37.65	41.57	44.31
26	12.20	13.41	15.38	17.29	25.34	35.56	38.88	42.86	45.64
27	12.88	14.13	16.15	18.11	26.34	36.74	40.11	44.14	46.96
28	13.57	14.85	16.93	18.94	27.34	37.92	41.34	45.42	48.28
29	14.26	15.57	17.71	19.77	28.34	39.09	42.56	46.69	49.59
30	14.95	16.31	18.49	20.60	29.34	40.26	43.77	47.96	50.89
n P	.99	.98	.95	.90	.50	.10	.05	.02	.01

For $n > 30$, let $z = \sqrt{2\chi^2} - \sqrt{2n - 1}$, and use the normal probability function, $\phi(z)$.
Table F is abridged from Table III of Fisher: *Statistical Methods for Research Workers*, Oliver & Boyd, Ltd., Edinburgh, by permission of the author and publishers.

Table G(a)—CRITICAL VALUES OF F

$P = 0.05$ (Roman Type) and $P = 0.01$ (Bold Face Type) are the probabilities of exceeding the tabulated values of F corresponding to n_1 and n_2 degrees of freedom where n_1 is associated with the greater variance.

$n_2 \backslash n_1$	1	2	3	4	5	6	7	8	9	10	11	12	14	16	20	24	30	40	50	75	100	200	500	∞
1	161 **4,052**	200 **4,999**	216 **5,403**	225 **5,625**	230 **5,764**	234 **5,859**	237 **5,928**	239 **5,981**	241 **6,022**	242 **6,056**	243 **6,082**	244 **6,106**	245 **6,142**	246 **6,169**	248 **6,208**	249 **6,234**	250 **6,258**	251 **6,286**	252 **6,302**	253 **6,323**	253 **6,334**	254 **6,352**	254 **6,361**	254 **6,366**
2	18.51 **98.49**	19.00 **99.00**	19.16 **99.17**	19.25 **99.25**	19.30 **99.30**	19.33 **99.33**	19.36 **99.34**	19.37 **99.36**	19.38 **99.38**	19.39 **99.40**	19.40 **99.41**	19.41 **99.42**	19.42 **99.43**	19.43 **99.44**	19.44 **99.45**	19.45 **99.46**	19.46 **99.47**	19.47 **99.48**	19.47 **99.48**	19.48 **99.49**	19.49 **99.49**	19.49 **99.49**	19.50 **99.50**	19.50 **99.50**
3	10.13 **34.12**	9.55 **30.82**	9.28 **29.46**	9.12 **28.71**	9.01 **28.24**	8.94 **27.91**	8.88 **27.67**	8.84 **27.49**	8.81 **27.34**	8.78 **27.23**	8.76 **27.13**	8.74 **27.05**	8.71 **26.92**	8.69 **26.83**	8.66 **26.69**	8.64 **26.60**	8.62 **26.50**	8.60 **26.41**	8.58 **26.35**	8.57 **26.27**	8.56 **26.23**	8.54 **26.18**	8.54 **26.14**	8.53 **26.12**
4	7.71 **21.20**	6.94 **18.00**	6.59 **16.69**	6.39 **15.98**	6.26 **15.52**	6.16 **15.21**	6.09 **14.98**	6.04 **14.80**	6.00 **14.66**	5.96 **14.54**	5.93 **14.45**	5.91 **14.37**	5.87 **14.24**	5.84 **14.15**	5.80 **14.02**	5.77 **13.93**	5.74 **13.83**	5.71 **13.74**	5.70 **13.69**	5.68 **13.61**	5.66 **13.57**	5.65 **13.52**	5.64 **13.48**	5.63 **13.46**
5	6.61 **16.26**	5.79 **13.27**	5.41 **12.06**	5.19 **11.39**	5.05 **10.97**	4.95 **10.67**	4.88 **10.45**	4.82 **10.27**	4.78 **10.15**	4.74 **10.05**	4.70 **9.96**	4.68 **9.89**	4.64 **9.77**	4.60 **9.68**	4.56 **9.55**	4.53 **9.47**	4.50 **9.38**	4.46 **9.29**	4.44 **9.24**	4.42 **9.17**	4.40 **9.13**	4.38 **9.07**	4.37 **9.04**	4.36 **9.02**
6	5.99 **13.74**	5.14 **10.92**	4.76 **9.78**	4.53 **9.15**	4.39 **8.75**	4.28 **8.47**	4.21 **8.26**	4.15 **8.10**	4.10 **7.98**	4.06 **7.87**	4.03 **7.79**	4.00 **7.72**	3.96 **7.60**	3.92 **7.52**	3.87 **7.39**	3.84 **7.31**	3.81 **7.23**	3.77 **7.14**	3.75 **7.09**	3.72 **7.02**	3.71 **6.99**	3.69 **6.94**	3.68 **6.90**	3.67 **6.88**
7	5.59 **12.25**	4.74 **9.55**	4.35 **8.45**	4.12 **7.85**	3.97 **7.46**	3.87 **7.19**	3.79 **7.00**	3.73 **6.84**	3.68 **6.71**	3.63 **6.62**	3.60 **6.54**	3.57 **6.47**	3.52 **6.35**	3.49 **6.27**	3.44 **6.15**	3.41 **6.07**	3.38 **5.98**	3.34 **5.90**	3.32 **5.85**	3.29 **5.78**	3.28 **5.75**	3.25 **5.70**	3.24 **5.67**	3.23 **5.65**
8	5.32 **11.26**	4.46 **8.65**	4.07 **7.59**	3.84 **7.01**	3.69 **6.63**	3.58 **6.37**	3.50 **6.19**	3.44 **6.03**	3.39 **5.91**	3.34 **5.82**	3.31 **5.74**	3.28 **5.67**	3.23 **5.56**	3.20 **5.48**	3.15 **5.36**	3.12 **5.28**	3.08 **5.20**	3.05 **5.11**	3.03 **5.06**	3.00 **5.00**	2.98 **4.96**	2.96 **4.91**	2.94 **4.88**	2.93 **4.86**
9	5.12 **10.56**	4.26 **8.02**	3.86 **6.99**	3.63 **6.42**	3.48 **6.06**	3.37 **5.80**	3.29 **5.62**	3.23 **5.47**	3.18 **5.35**	3.13 **5.26**	3.10 **5.18**	3.07 **5.11**	3.02 **5.00**	2.98 **4.92**	2.93 **4.80**	2.90 **4.73**	2.86 **4.64**	2.82 **4.56**	2.80 **4.51**	2.77 **4.45**	2.76 **4.41**	2.73 **4.36**	2.72 **4.33**	2.71 **4.31**
10	4.96 **10.04**	4.10 **7.56**	3.71 **6.55**	3.48 **5.99**	3.33 **5.64**	3.22 **5.39**	3.14 **5.21**	3.07 **5.06**	3.02 **4.95**	2.97 **4.85**	2.94 **4.78**	2.91 **4.71**	2.86 **4.60**	2.82 **4.52**	2.77 **4.41**	2.74 **4.33**	2.70 **4.25**	2.67 **4.17**	2.64 **4.12**	2.61 **4.05**	2.59 **4.01**	2.56 **3.96**	2.55 **3.93**	2.54 **3.91**
11	4.84 **9.65**	3.98 **7.20**	3.59 **6.22**	3.36 **5.67**	3.20 **5.32**	3.09 **5.07**	3.01 **4.88**	2.95 **4.74**	2.90 **4.63**	2.86 **4.54**	2.82 **4.46**	2.79 **4.40**	2.74 **4.29**	2.70 **4.21**	2.65 **4.10**	2.61 **4.02**	2.57 **3.94**	2.53 **3.86**	2.50 **3.80**	2.47 **3.74**	2.45 **3.70**	2.42 **3.66**	2.41 **3.62**	2.40 **3.60**
12	4.75 **9.33**	3.88 **6.93**	3.49 **5.95**	3.26 **5.41**	3.11 **5.06**	3.00 **4.82**	2.92 **4.65**	2.85 **4.50**	2.80 **4.39**	2.76 **4.30**	2.72 **4.22**	2.69 **4.16**	2.64 **4.05**	2.60 **3.98**	2.54 **3.86**	2.50 **3.78**	2.46 **3.70**	2.42 **3.61**	2.40 **3.56**	2.36 **3.49**	2.35 **3.46**	2.32 **3.41**	2.31 **3.38**	2.30 **3.36**
13	4.67 **9.07**	3.80 **6.70**	3.41 **5.74**	3.18 **5.20**	3.02 **4.86**	2.92 **4.62**	2.84 **4.44**	2.77 **4.30**	2.72 **4.19**	2.67 **4.10**	2.63 **4.02**	2.60 **3.96**	2.55 **3.85**	2.51 **3.78**	2.46 **3.67**	2.42 **3.59**	2.38 **3.51**	2.34 **3.42**	2.32 **3.37**	2.28 **3.30**	2.26 **3.27**	2.24 **3.21**	2.22 **3.18**	2.21 **3.16**

Table G(a) (Continued)

n_2 \ n_1	1	2	3	4	5	6	7	8	9	10	11	12	14	16	20	24	30	40	50	75	100	200	500	∞
14	4.60 / 8.86	3.74 / 6.51	3.34 / 5.56	3.11 / 5.03	2.96 / 4.69	2.85 / 4.46	2.77 / 4.28	2.70 / 4.14	2.65 / 4.03	2.60 / 3.94	2.56 / 3.86	2.53 / 3.80	2.48 / 3.70	2.44 / 3.62	2.39 / 3.51	2.35 / 3.43	2.31 / 3.34	2.27 / 3.26	2.24 / 3.21	2.21 / 3.14	2.19 / 3.11	2.16 / 3.06	2.14 / 3.02	2.13 / 3.00
15	4.54 / 8.68	3.68 / 6.36	3.29 / 5.42	3.06 / 4.89	2.90 / 4.56	2.79 / 4.32	2.70 / 4.14	2.64 / 4.00	2.59 / 3.89	2.55 / 3.80	2.51 / 3.73	2.48 / 3.67	2.43 / 3.56	2.39 / 3.48	2.33 / 3.36	2.29 / 3.29	2.25 / 3.20	2.21 / 3.12	2.18 / 3.07	2.15 / 3.00	2.12 / 2.97	2.10 / 2.92	2.08 / 2.89	2.07 / 2.87
16	4.49 / 8.53	3.63 / 6.23	3.24 / 5.29	3.01 / 4.77	2.85 / 4.44	2.74 / 4.20	2.66 / 4.03	2.59 / 3.89	2.54 / 3.78	2.49 / 3.69	2.45 / 3.61	2.42 / 3.55	2.37 / 3.45	2.33 / 3.37	2.28 / 3.25	2.24 / 3.18	2.20 / 3.10	2.16 / 3.01	2.13 / 2.96	2.09 / 2.89	2.07 / 2.86	2.04 / 2.80	2.02 / 2.77	2.01 / 2.75
17	4.45 / 8.40	3.59 / 6.11	3.20 / 5.18	2.96 / 4.67	2.81 / 4.34	2.70 / 4.10	2.62 / 3.93	2.55 / 3.79	2.50 / 3.68	2.45 / 3.59	2.41 / 3.52	2.38 / 3.45	2.33 / 3.35	2.29 / 3.27	2.23 / 3.16	2.19 / 3.08	2.15 / 3.00	2.11 / 2.92	2.08 / 2.86	2.04 / 2.79	2.02 / 2.76	1.99 / 2.70	1.97 / 2.67	1.96 / 2.65
18	4.41 / 8.28	3.55 / 6.01	3.13 / 5.09	2.93 / 4.58	2.77 / 4.25	2.66 / 4.01	2.58 / 3.85	2.51 / 3.71	2.46 / 3.60	2.41 / 3.51	2.37 / 3.44	2.34 / 3.37	2.29 / 3.27	2.25 / 3.19	2.19 / 3.07	2.15 / 3.00	2.11 / 2.91	2.07 / 2.83	2.04 / 2.78	2.00 / 2.71	1.98 / 2.68	1.95 / 2.62	1.93 / 2.59	1.92 / 2.57
19	4.38 / 8.18	3.52 / 5.93	3.13 / 5.01	2.90 / 4.50	2.74 / 4.17	2.63 / 3.94	2.55 / 3.77	2.48 / 3.63	2.43 / 3.52	2.38 / 3.43	2.34 / 3.36	2.31 / 3.30	2.26 / 3.19	2.21 / 3.12	2.15 / 3.00	2.11 / 2.92	2.07 / 2.84	2.02 / 2.76	2.00 / 2.70	1.96 / 2.63	1.94 / 2.60	1.91 / 2.54	1.90 / 2.51	1.88 / 2.49
20	4.35 / 8.10	3.49 / 5.85	3.10 / 4.94	2.87 / 4.43	2.71 / 4.10	2.60 / 3.87	2.52 / 3.71	2.45 / 3.56	2.40 / 3.45	2.35 / 3.37	2.31 / 3.30	2.28 / 3.23	2.23 / 3.13	2.18 / 3.05	2.12 / 2.94	2.08 / 2.86	2.04 / 2.77	1.99 / 2.69	1.96 / 2.63	1.92 / 2.56	1.90 / 2.53	1.87 / 2.47	1.85 / 2.44	1.84 / 2.42
21	4.32 / 8.02	3.47 / 5.78	3.07 / 4.87	2.84 / 4.37	2.68 / 4.04	2.57 / 3.81	2.49 / 3.65	2.42 / 3.51	2.37 / 3.40	2.32 / 3.31	2.28 / 3.24	2.25 / 3.17	2.20 / 3.07	2.15 / 2.99	2.09 / 2.88	2.05 / 2.80	2.00 / 2.72	1.96 / 2.63	1.93 / 2.58	1.89 / 2.51	1.87 / 2.47	1.84 / 2.42	1.82 / 2.38	1.81 / 2.36
22	4.30 / 7.94	3.44 / 5.72	3.05 / 4.82	2.82 / 4.31	2.66 / 3.99	2.55 / 3.76	2.47 / 3.59	2.40 / 3.45	2.35 / 3.35	2.30 / 3.26	2.26 / 3.18	2.23 / 3.12	2.18 / 3.02	2.13 / 2.94	2.07 / 2.83	2.03 / 2.75	1.98 / 2.67	1.93 / 2.58	1.91 / 2.53	1.87 / 2.46	1.84 / 2.42	1.81 / 2.37	1.80 / 2.33	1.78 / 2.31
23	4.28 / 7.88	3.42 / 5.66	3.03 / 4.76	2.80 / 4.26	2.64 / 3.94	2.53 / 3.71	2.45 / 3.54	2.38 / 3.41	2.32 / 3.30	2.28 / 3.21	2.24 / 3.14	2.20 / 3.07	2.14 / 2.97	2.10 / 2.89	2.04 / 2.78	2.00 / 2.70	1.96 / 2.62	1.91 / 2.53	1.88 / 2.48	1.84 / 2.41	1.82 / 2.37	1.79 / 2.32	1.77 / 2.28	1.76 / 2.26
24	4.26 / 7.82	3.40 / 5.61	3.01 / 4.72	2.78 / 4.22	2.62 / 3.90	2.51 / 3.67	2.43 / 3.50	2.36 / 3.36	2.30 / 3.25	2.26 / 3.17	2.22 / 3.09	2.18 / 3.03	2.13 / 2.93	2.09 / 2.85	2.02 / 2.74	1.98 / 2.66	1.94 / 2.58	1.89 / 2.49	1.86 / 2.44	1.82 / 2.36	1.80 / 2.33	1.76 / 2.27	1.74 / 2.23	1.73 / 2.21
25	4.24 / 7.77	3.38 / 5.57	2.99 / 4.68	2.76 / 4.18	2.60 / 3.86	2.49 / 3.63	2.41 / 3.46	2.34 / 3.32	2.28 / 3.21	2.24 / 3.13	2.20 / 3.05	2.16 / 2.99	2.11 / 2.89	2.06 / 2.81	2.00 / 2.70	1.96 / 2.62	1.92 / 2.54	1.87 / 2.45	1.84 / 2.40	1.80 / 2.32	1.77 / 2.29	1.74 / 2.23	1.72 / 2.19	1.71 / 2.17
26	4.22 / 7.72	3.37 / 5.53	2.98 / 4.64	2.74 / 4.14	2.59 / 3.82	2.47 / 3.59	2.39 / 3.42	2.32 / 3.29	2.27 / 3.17	2.22 / 3.09	2.18 / 3.02	2.15 / 2.96	2.10 / 2.86	2.05 / 2.77	1.99 / 2.66	1.95 / 2.58	1.90 / 2.50	1.85 / 2.41	1.82 / 2.36	1.78 / 2.28	1.76 / 2.25	1.72 / 2.19	1.70 / 2.15	1.69 / 2.13

The function, $F = e$ with exponent $2z$, is computed in part from Fisher's Table VI (7). Additional entries are by interpolation, mostly graphical.
Table G(a) is reproduced from Snedecor: *Statistical Methods*, Iowa State College Press, Ames, Iowa, with the kind permission of the author and publishers.

Table G(a) (Continued)

n_1 / n_2	1	2	3	4	5	6	7	8	9	10	11	12	14	16	20	24	30	40	50	75	100	200	500	∞
27	4.21 **7.68**	3.35 **5.49**	2.96 **4.60**	2.73 **4.11**	2.57 **3.79**	2.46 **3.56**	2.37 **3.39**	2.30 **3.26**	2.25 **3.14**	2.20 **3.06**	2.16 **2.98**	2.13 **2.93**	2.08 **2.83**	2.03 **2.74**	1.97 **2.63**	1.93 **2.55**	1.88 **2.47**	1.84 **2.38**	1.80 **2.33**	1.76 **2.25**	1.74 **2.21**	1.71 **2.16**	1.68 **2.12**	1.67 **2.10**
28	4.20 **7.64**	3.34 **5.45**	2.95 **4.57**	2.71 **4.07**	2.56 **3.76**	2.44 **3.53**	2.36 **3.36**	2.29 **3.23**	2.24 **3.11**	2.19 **3.03**	2.15 **2.95**	2.12 **2.90**	2.06 **2.80**	2.02 **2.71**	1.96 **2.60**	1.91 **2.52**	1.87 **2.44**	1.81 **2.35**	1.78 **2.30**	1.75 **2.22**	1.72 **2.18**	1.69 **2.13**	1.67 **2.09**	1.65 **2.06**
29	4.18 **7.60**	3.33 **5.42**	2.93 **4.54**	2.70 **4.04**	2.54 **3.73**	2.43 **3.50**	2.35 **3.33**	2.28 **3.20**	2.22 **3.08**	2.18 **3.00**	2.14 **2.92**	2.10 **2.87**	2.05 **2.77**	2.00 **2.68**	1.94 **2.57**	1.90 **2.49**	1.85 **2.41**	1.80 **2.32**	1.77 **2.27**	1.73 **2.19**	1.71 **2.15**	1.68 **2.10**	1.65 **2.06**	1.64 **2.03**
30	4.17 **7.56**	3.32 **5.39**	2.92 **4.51**	2.69 **4.02**	2.53 **3.70**	2.42 **3.47**	2.34 **3.30**	2.27 **3.17**	2.21 **3.06**	2.16 **2.98**	2.12 **2.90**	2.09 **2.84**	2.04 **2.74**	1.99 **2.66**	1.93 **2.55**	1.89 **2.47**	1.84 **2.38**	1.79 **2.29**	1.76 **2.24**	1.72 **2.16**	1.69 **2.13**	1.66 **2.07**	1.64 **2.03**	1.62 **2.01**
32	4.15 **7.50**	3.30 **5.34**	2.90 **4.46**	2.67 **3.97**	2.51 **3.66**	2.40 **3.42**	2.32 **3.25**	2.25 **3.12**	2.19 **3.01**	2.14 **2.94**	2.10 **2.86**	2.07 **2.80**	2.02 **2.70**	1.97 **2.62**	1.91 **2.51**	1.86 **2.42**	1.82 **2.34**	1.76 **2.25**	1.74 **2.20**	1.69 **2.12**	1.67 **2.08**	1.64 **2.02**	1.61 **1.98**	1.59 **1.96**
34	4.13 **7.44**	3.28 **5.29**	2.88 **4.42**	2.65 **3.93**	2.49 **3.61**	2.38 **3.38**	2.30 **3.21**	2.23 **3.08**	2.17 **2.97**	2.12 **2.89**	2.08 **2.82**	2.05 **2.76**	2.00 **2.66**	1.95 **2.58**	1.89 **2.47**	1.84 **2.38**	1.80 **2.30**	1.74 **2.21**	1.71 **2.15**	1.67 **2.08**	1.64 **2.04**	1.61 **1.98**	1.59 **1.94**	1.57 **1.91**
36	4.11 **7.39**	3.26 **5.25**	2.86 **4.38**	2.63 **3.89**	2.48 **3.58**	2.36 **3.35**	2.28 **3.18**	2.21 **3.04**	2.15 **2.94**	2.10 **2.86**	2.06 **2.78**	2.03 **2.72**	1.98 **2.62**	1.93 **2.54**	1.87 **2.43**	1.82 **2.35**	1.78 **2.26**	1.72 **2.17**	1.69 **2.12**	1.65 **2.04**	1.62 **2.00**	1.59 **1.94**	1.56 **1.90**	1.55 **1.87**
38	4.10 **7.35**	3.25 **5.21**	2.85 **4.34**	2.62 **3.86**	2.46 **3.54**	2.35 **3.32**	2.26 **3.15**	2.19 **3.02**	2.14 **2.91**	2.09 **2.82**	2.05 **2.75**	2.02 **2.69**	1.96 **2.59**	1.92 **2.51**	1.85 **2.40**	1.80 **2.32**	1.76 **2.22**	1.71 **2.14**	1.67 **2.08**	1.63 **2.00**	1.60 **1.97**	1.57 **1.90**	1.54 **1.86**	1.53 **1.84**
40	4.08 **7.31**	3.23 **5.18**	2.84 **4.31**	2.61 **3.83**	2.45 **3.51**	2.34 **3.29**	2.25 **3.12**	2.18 **2.99**	2.12 **2.88**	2.07 **2.80**	2.04 **2.73**	2.00 **2.66**	1.95 **2.56**	1.90 **2.49**	1.84 **2.37**	1.79 **2.29**	1.74 **2.20**	1.69 **2.11**	1.66 **2.05**	1.61 **1.97**	1.59 **1.94**	1.55 **1.88**	1.53 **1.84**	1.51 **1.81**
42	4.07 **7.27**	3.22 **5.15**	2.83 **4.29**	2.59 **3.80**	2.44 **3.49**	2.32 **3.26**	2.24 **3.10**	2.17 **2.96**	2.11 **2.86**	2.06 **2.77**	2.02 **2.70**	1.99 **2.64**	1.94 **2.54**	1.89 **2.46**	1.82 **2.35**	1.78 **2.26**	1.73 **2.17**	1.68 **2.08**	1.64 **2.02**	1.60 **1.94**	1.57 **1.91**	1.54 **1.85**	1.51 **1.80**	1.49 **1.78**
44	4.06 **7.24**	3.21 **5.12**	2.82 **4.26**	2.58 **3.78**	2.43 **3.46**	2.31 **3.24**	2.23 **3.07**	2.16 **2.94**	2.10 **2.84**	2.05 **2.75**	2.01 **2.68**	1.98 **2.62**	1.92 **2.52**	1.88 **2.44**	1.81 **2.32**	1.76 **2.24**	1.72 **2.15**	1.66 **2.06**	1.63 **2.00**	1.58 **1.92**	1.56 **1.88**	1.52 **1.82**	1.50 **1.78**	1.48 **1.75**
46	4.05 **7.21**	3.20 **5.10**	2.81 **4.24**	2.57 **3.76**	2.42 **3.44**	2.30 **3.22**	2.22 **3.05**	2.14 **2.92**	2.09 **2.82**	2.04 **2.73**	2.00 **2.66**	1.97 **2.60**	1.91 **2.50**	1.87 **2.42**	1.80 **2.30**	1.75 **2.22**	1.71 **2.13**	1.65 **2.04**	1.62 **1.98**	1.57 **1.90**	1.54 **1.86**	1.51 **1.80**	1.48 **1.76**	1.46 **1.72**
48	4.04 **7.19**	3.19 **5.08**	2.80 **4.22**	2.56 **3.74**	2.41 **3.42**	2.30 **3.20**	2.21 **3.04**	2.14 **2.90**	2.08 **2.80**	2.03 **2.71**	1.99 **2.64**	1.96 **2.58**	1.90 **2.48**	1.86 **2.40**	1.79 **2.28**	1.74 **2.20**	1.70 **2.11**	1.64 **2.02**	1.61 **1.96**	1.56 **1.88**	1.53 **1.84**	1.50 **1.78**	1.47 **1.73**	1.45 **1.70**

302

n_2 \ n_1	1	2	3	4	5	6	7	8	9	10	11	12	14	15	20	24	30	40	50	75	100	200	500	∞
50	4.03 / **7.17**	3.18 / **5.06**	2.79 / **4.20**	2.56 / **3.72**	2.40 / **3.41**	2.29 / **3.18**	2.20 / **3.02**	2.13 / **2.88**	2.07 / **2.78**	2.02 / **2.70**	1.98 / **2.62**	1.95 / **2.56**	1.90 / **2.46**	1.85 / **2.39**	1.78 / **2.26**	1.74 / **2.18**	1.69 / **2.10**	1.63 / **2.00**	1.60 / **1.94**	1.55 / **1.86**	1.52 / **1.82**	1.48 / **1.76**	1.46 / **1.71**	1.44 / **1.68**
55	4.02 / **7.12**	3.17 / **5.01**	2.78 / **4.16**	2.54 / **3.68**	2.38 / **3.37**	2.27 / **3.15**	2.18 / **2.98**	2.11 / **2.85**	2.05 / **2.75**	2.00 / **2.66**	1.97 / **2.59**	1.93 / **2.53**	1.88 / **2.43**	1.83 / **2.33**	1.76 / **2.23**	1.72 / **2.15**	1.67 / **2.06**	1.61 / **1.96**	1.58 / **1.90**	1.52 / **1.82**	1.50 / **1.78**	1.46 / **1.71**	1.43 / **1.66**	1.41 / **1.64**
60	4.00 / **7.08**	3.15 / **4.98**	2.76 / **4.13**	2.52 / **3.65**	2.37 / **3.34**	2.25 / **3.12**	2.17 / **2.95**	2.10 / **2.82**	2.04 / **2.72**	1.99 / **2.63**	1.95 / **2.56**	1.92 / **2.50**	1.86 / **2.40**	1.81 / **2.31**	1.75 / **2.20**	1.70 / **2.12**	1.65 / **2.03**	1.59 / **1.93**	1.56 / **1.87**	1.50 / **1.79**	1.48 / **1.74**	1.44 / **1.68**	1.41 / **1.63**	1.39 / **1.60**
65	3.99 / **7.04**	3.14 / **4.95**	2.75 / **4.10**	2.51 / **3.62**	2.36 / **3.31**	2.24 / **3.09**	2.15 / **2.93**	2.08 / **2.79**	2.02 / **2.70**	1.98 / **2.61**	1.94 / **2.54**	1.90 / **2.47**	1.85 / **2.37**	1.80 / **2.30**	1.73 / **2.18**	1.68 / **2.09**	1.63 / **2.00**	1.57 / **1.90**	1.54 / **1.84**	1.49 / **1.76**	1.46 / **1.71**	1.42 / **1.64**	1.39 / **1.60**	1.37 / **1.56**
70	3.98 / **7.01**	3.13 / **4.92**	2.74 / **4.08**	2.50 / **3.60**	2.35 / **3.29**	2.23 / **3.07**	2.14 / **2.91**	2.07 / **2.77**	2.01 / **2.67**	1.97 / **2.59**	1.93 / **2.51**	1.89 / **2.45**	1.84 / **2.35**	1.79 / **2.28**	1.72 / **2.15**	1.67 / **2.07**	1.62 / **1.98**	1.56 / **1.88**	1.53 / **1.82**	1.47 / **1.74**	1.45 / **1.69**	1.40 / **1.62**	1.37 / **1.56**	1.35 / **1.53**
80	3.96 / **6.96**	3.11 / **4.88**	2.72 / **4.04**	2.48 / **3.56**	2.33 / **3.25**	2.21 / **3.04**	2.12 / **2.87**	2.05 / **2.74**	1.99 / **2.64**	1.95 / **2.55**	1.91 / **2.48**	1.88 / **2.41**	1.82 / **2.32**	1.77 / **2.24**	1.70 / **2.11**	1.65 / **2.03**	1.60 / **1.94**	1.54 / **1.84**	1.51 / **1.78**	1.45 / **1.70**	1.42 / **1.65**	1.38 / **1.57**	1.35 / **1.52**	1.32 / **1.49**
100	3.94 / **6.90**	3.09 / **4.82**	2.70 / **3.98**	2.46 / **3.51**	2.30 / **3.20**	2.19 / **2.99**	2.10 / **2.82**	2.03 / **2.69**	1.97 / **2.59**	1.92 / **2.51**	1.88 / **2.43**	1.85 / **2.36**	1.79 / **2.26**	1.75 / **2.19**	1.68 / **2.06**	1.63 / **1.98**	1.57 / **1.89**	1.51 / **1.79**	1.48 / **1.73**	1.42 / **1.64**	1.39 / **1.59**	1.34 / **1.51**	1.30 / **1.46**	1.28 / **1.43**
125	3.92 / **6.84**	3.07 / **4.78**	2.68 / **3.94**	2.44 / **3.47**	2.29 / **3.17**	2.17 / **2.95**	2.08 / **2.79**	2.01 / **2.65**	1.95 / **2.56**	1.90 / **2.47**	1.86 / **2.40**	1.83 / **2.33**	1.77 / **2.23**	1.72 / **2.15**	1.65 / **2.03**	1.60 / **1.94**	1.55 / **1.85**	1.49 / **1.75**	1.45 / **1.68**	1.39 / **1.59**	1.36 / **1.54**	1.31 / **1.46**	1.27 / **1.40**	1.25 / **1.37**
150	3.91 / **6.81**	3.06 / **4.75**	2.67 / **3.91**	2.43 / **3.44**	2.27 / **3.14**	2.16 / **2.92**	2.07 / **2.76**	2.00 / **2.62**	1.94 / **2.53**	1.89 / **2.44**	1.85 / **2.37**	1.82 / **2.30**	1.76 / **2.20**	1.71 / **2.12**	1.64 / **2.00**	1.59 / **1.91**	1.54 / **1.83**	1.47 / **1.72**	1.44 / **1.66**	1.37 / **1.56**	1.34 / **1.51**	1.29 / **1.43**	1.25 / **1.37**	1.22 / **1.33**
200	3.89 / **6.76**	3.04 / **4.71**	2.65 / **3.88**	2.41 / **3.41**	2.26 / **3.11**	2.14 / **2.90**	2.05 / **2.73**	1.98 / **2.60**	1.92 / **2.50**	1.87 / **2.41**	1.83 / **2.34**	1.80 / **2.28**	1.74 / **2.17**	1.69 / **2.09**	1.62 / **1.97**	1.57 / **1.88**	1.52 / **1.79**	1.45 / **1.69**	1.42 / **1.62**	1.35 / **1.53**	1.32 / **1.48**	1.26 / **1.39**	1.22 / **1.33**	1.19 / **1.28**
400	3.86 / **6.70**	3.02 / **4.66**	2.62 / **3.83**	2.39 / **3.36**	2.23 / **3.06**	2.12 / **2.85**	2.03 / **2.69**	1.96 / **2.55**	1.90 / **2.46**	1.85 / **2.37**	1.81 / **2.29**	1.78 / **2.23**	1.72 / **2.12**	1.67 / **2.04**	1.60 / **1.92**	1.54 / **1.84**	1.49 / **1.74**	1.42 / **1.64**	1.38 / **1.57**	1.32 / **1.47**	1.28 / **1.42**	1.22 / **1.32**	1.16 / **1.24**	1.13 / **1.19**
1000	3.85 / **6.66**	3.00 / **4.62**	2.61 / **3.80**	2.38 / **3.34**	2.22 / **3.04**	2.10 / **2.82**	2.02 / **2.66**	1.95 / **2.53**	1.89 / **2.43**	1.84 / **2.34**	1.80 / **2.26**	1.76 / **2.20**	1.70 / **2.09**	1.65 / **2.01**	1.58 / **1.89**	1.53 / **1.81**	1.47 / **1.71**	1.41 / **1.61**	1.36 / **1.54**	1.30 / **1.44**	1.26 / **1.38**	1.19 / **1.28**	1.13 / **1.19**	1.08 / **1.11**
∞	3.84 / **6.64**	2.99 / **4.60**	2.60 / **3.78**	2.37 / **3.32**	2.21 / **3.02**	2.09 / **2.80**	2.01 / **2.64**	1.94 / **2.51**	1.88 / **2.41**	1.83 / **2.32**	1.79 / **2.24**	1.75 / **2.18**	1.69 / **2.07**	1.64 / **1.99**	1.57 / **1.87**	1.52 / **1.79**	1.46 / **1.69**	1.40 / **1.59**	1.35 / **1.52**	1.28 / **1.41**	1.24 / **1.36**	1.17 / **1.25**	1.11 / **1.15**	1.00 / **1.00**

Table G(b)—Critical Values Of F

$P = 0.025$ is the probability of exceeding the tabulated value of F corresponding to n_1 and n_2 degrees of freedom, where n_1 is associated with the greater variance.

$n_2 \backslash n_1$	1	2	3	4	5	6	7	8	9	10	12	15	20	24	30	40	60	120	∞
1	647.8	799.5	864.2	899.6	921.8	937.1	948.2	956.7	963.3	968.6	976.7	984.9	993.1	997.2	1001	1006	1010	1014	1018
2	38.51	39.00	39.17	39.25	39.30	39.33	39.36	39.37	39.39	39.40	39.41	39.43	39.45	39.46	39.46	39.47	39.48	39.49	39.50
3	17.44	16.04	15.44	15.10	14.88	14.73	14.62	14.54	14.47	14.42	14.34	14.25	14.17	14.12	14.08	14.04	13.99	13.95	13.90
4	12.22	10.65	9.98	9.60	9.36	9.20	9.07	8.98	8.90	8.84	8.75	8.66	8.56	8.51	8.46	8.41	8.36	8.31	8.26
5	10.01	8.43	7.76	7.39	7.15	6.98	6.85	6.76	6.68	6.62	6.52	6.43	6.33	6.28	6.23	6.18	6.12	6.07	6.02
6	8.81	7.26	6.60	6.23	5.99	5.82	5.70	5.60	5.52	5.46	5.37	5.27	5.17	5.12	5.07	5.01	4.96	4.90	4.85
7	8.07	6.54	5.89	5.52	5.29	5.12	4.99	4.90	4.82	4.76	4.67	4.57	4.47	4.42	4.36	4.31	4.25	4.20	4.14
8	7.57	6.06	5.42	5.05	4.82	4.65	4.53	4.43	4.36	4.30	4.20	4.10	4.00	3.95	3.89	3.84	3.78	3.73	3.67
9	7.21	5.71	5.08	4.72	4.48	4.32	4.20	4.10	4.03	3.96	3.87	3.77	3.67	3.61	3.56	3.51	3.45	3.39	3.33
10	6.94	5.46	4.83	4.47	4.24	4.07	3.95	3.85	3.78	3.72	3.62	3.52	3.42	3.37	3.31	3.26	3.20	3.14	3.08
11	6.72	5.26	4.63	4.28	4.04	3.88	3.76	3.66	3.59	3.53	3.43	3.33	3.23	3.17	3.12	3.06	3.00	2.94	2.88
12	6.55	5.10	4.47	4.12	3.89	3.73	3.61	3.51	3.44	3.37	3.28	3.18	3.07	3.02	2.96	2.91	2.85	2.79	2.72
13	6.41	4.97	4.35	4.00	3.77	3.60	3.48	3.39	3.31	3.25	3.15	3.05	2.95	2.89	2.84	2.78	2.72	2.66	2.60
14	6.30	4.86	4.24	3.89	3.66	3.50	3.38	3.29	3.21	3.15	3.05	2.95	2.84	2.79	2.73	2.67	2.61	2.55	2.49
15	6.20	4.77	4.15	3.80	3.58	3.41	3.29	3.20	3.12	3.06	2.96	2.86	2.76	2.70	2.64	2.59	2.52	2.46	2.40
16	6.12	4.69	4.08	3.73	3.50	3.34	3.22	3.12	3.05	2.99	2.89	2.79	2.68	2.63	2.57	2.51	2.45	2.38	2.32
17	6.04	4.62	4.01	3.66	3.44	3.28	3.16	3.06	2.98	2.92	2.82	2.72	2.62	2.56	2.50	2.44	2.38	2.32	2.25
18	5.98	4.56	3.95	3.61	3.38	3.22	3.10	3.01	2.93	2.87	2.77	2.67	2.56	2.50	2.44	2.38	2.32	2.26	2.19
19	5.92	4.51	3.90	3.56	3.33	3.17	3.05	2.96	2.88	2.82	2.72	2.62	2.51	2.45	2.39	2.33	2.27	2.20	2.13
20	5.87	4.46	3.86	3.51	3.29	3.13	3.01	2.91	2.84	2.77	2.68	2.57	2.46	2.41	2.35	2.29	2.22	2.16	2.09
21	5.83	4.42	3.82	3.48	3.25	3.09	2.97	2.87	2.80	2.73	2.64	2.53	2.42	2.37	2.31	2.25	2.18	2.11	2.04
22	5.79	4.38	3.78	3.44	3.22	3.05	2.93	2.84	2.76	2.70	2.60	2.50	2.39	2.33	2.27	2.21	2.14	2.08	2.00
23	5.75	4.35	3.75	3.41	3.18	3.02	2.90	2.81	2.73	2.67	2.57	2.47	2.36	2.30	2.24	2.18	2.11	2.04	1.97
24	5.72	4.32	3.72	3.38	3.15	2.99	2.87	2.78	2.70	2.64	2.54	2.44	2.33	2.27	2.21	2.15	2.08	2.01	1.94
25	5.69	4.29	3.69	3.35	3.13	2.97	2.85	2.75	2.68	2.61	2.51	2.41	2.30	2.24	2.18	2.12	2.05	1.98	1.91
26	5.66	4.27	3.67	3.33	3.10	2.94	2.82	2.73	2.65	2.59	2.49	2.39	2.28	2.22	2.16	2.09	2.03	1.95	1.88
27	5.63	4.24	3.65	3.31	3.08	2.92	2.80	2.71	2.63	2.57	2.47	2.36	2.25	2.19	2.13	2.07	2.00	1.93	1.85
28	5.61	4.22	3.63	3.29	3.06	2.90	2.78	2.69	2.61	2.55	2.45	2.34	2.23	2.17	2.11	2.05	1.98	1.91	1.83
29	5.59	4.20	3.61	3.27	3.04	2.88	2.76	2.67	2.59	2.53	2.43	2.32	2.21	2.15	2.09	2.03	1.96	1.89	1.81
30	5.57	4.18	3.59	3.25	3.03	2.87	2.75	2.65	2.57	2.51	2.41	2.31	2.20	2.14	2.07	2.01	1.94	1.87	1.79
40	5.42	4.05	3.46	3.13	2.90	2.74	2.62	2.53	2.45	2.39	2.29	2.18	2.07	2.01	1.94	1.88	1.80	1.72	1.64
60	5.29	3.93	3.34	3.01	2.79	2.63	2.51	2.41	2.33	2.27	2.17	2.06	1.94	1.88	1.82	1.74	1.67	1.58	1.48
120	5.15	3.80	3.23	2.89	2.67	2.52	2.39	2.30	2.22	2.16	2.05	1.94	1.82	1.76	1.69	1.61	1.53	1.43	1.31
∞	5.02	3.69	3.12	2.79	2.57	2.41	2.29	2.19	2.11	2.05	1.94	1.83	1.71	1.64	1.57	1.48	1.39	1.27	1.00

Table H—95% CONFIDENCE LIMITS FOR p

Each number attached to a curve denotes the sample size, N.

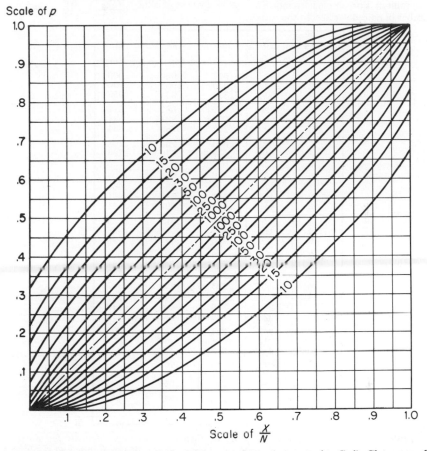

Scale of p

Scale of $\frac{X}{N}$

Reproduced by permission of E. S. Pearson from the paper by C. J. Clopper and E. S. Pearson, "The Use of Confidence or Fiducial Limits Illustrated in the Case of the Binomial," *Biometrika*, Vol. 26 (1934), p. 410.

Table I—Critical Values of r when $\rho = 0$

P equals the probability of exceeding numerically a given value of r corresponding to n degrees of freedom. (Two-tailed probability.) For simple correlation $n = N - 2$ where N is the number of paired variates in the sample.

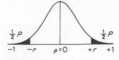

n \ P	.10	.05	.02	.01
1	.988	.997	1.000	1.000
2	.900	.950	.980	.990
3	.805	.878	.934	.959
4	.729	.811	.882	.917
5	.669	.754	.833	.874
6	.622	.707	.789	.834
7	.582	.666	.750	.798
8	.549	.632	.716	.765
9	.521	.602	.685	.735
10	.497	.576	.658	.708
11	.476	.553	.634	.684
12	.458	.532	.612	.661
13	.441	.514	.592	.641
14	.426	.497	.574	.623
15	.412	.482	.558	.606
16	.400	.468	.542	.590
17	.389	.456	.528	.575
18	.378	.444	.516	.561
19	.369	.433	.503	.549
20	.360	.423	.492	.537
25	.323	.381	.445	.487
30	.296	.349	.409	.449
35	.275	.325	.381	.418
40	.257	.304	.358	.393
45	.243	.288	.338	.372
50	.231	.273	.322	.354
60	.211	.250	.295	.325
70	.195	.232	.274	.302
80	.183	.217	.256	.283
90	.173	.205	.242	.267
100	.164	.195	.230	.254

Table I is reprinted from Table V.A. of Fisher: *Statistical Methods for Research Workers*, Oliver and Boyd, Ltd., Edinburgh, by permission of the author and publishers.

Table J—Values of Fisher's z Corresponding to Values of r

$$z = \tfrac{1}{2} \log_e \frac{1 + r}{1 - r}$$

r	.00	.01	.02	.03	.04	.05	.06	.07	.08	.09
.0	.000	.010	.020	.030	.040	.050	.060	.070	.080	.090
.1	.100	.110	.121	.131	.141	.151	.161	.172	.181	.192
.2	.203	.214	.224	.234	.245	.256	.266	.277	.288	.299
.3	.309	.321	.332	.343	.354	.366	.377	.389	.400	.412
.4	.424	.436	.448	.460	.472	.485	.497	.510	.523	.536
.5	.549	.563	.577	.590	.604	.618	.633	.648	.663	.678
.6	.693	.709	.725	.741	.758	.775	.793	.811	.829	.848
.7	.867	.887	.908	.929	.950	.973	.996	1.020	1.045	1.071
.8	1.099	1.127	1.157	1.188	1.221	1.256	1.293	1.333	1.376	1.422
.9	1.472	1.528	1.589	1.658	1.738	1.832	1.946	2.092	2.298	2.647

Table K—95% Confidence Limits for ρ

Each number attached to a curve denotes the sample size, N.

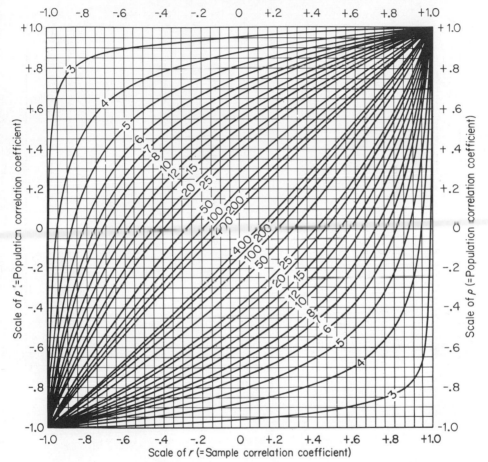

Scale of r (=Sample correlation coefficient)

MATHEMATICAL TABLES

Table L—CRITICAL VALUES FOR THE SIGN TEST

For N differences, where x is the number of occurrences of the less frequent sign, the column values under the headings 0.005, 0.025, 0.05, yield the values of r such that $P(x \leqq r) = 0.005, 0.025, 0.05$, respectively. (See Section 15.4.)

N	0.005	0.025	0.05	N	0.005	0.025	0.05
6		0	0	50	15	17	18
7		0	0				
8	0	0	1	51	15	18	19
9	0	1	1	52	16	18	19
				53	16	18	20
10	0	1	1				
				54	17	19	20
11	0	1	2	55	17	19	20
12	1	2	2	56	17	20	21
13	1	2	3				
				57	18	20	21
14	1	2	3	58	18	21	22
15	2	3	3	59	19	21	22
16	2	3	4				
				60	19	21	23
17	2	4	4				
18	3	4	5	61	20	22	23
19	3	4	5	62	20	22	24
				63	20	23	24
20	3	5	5				
				64	21	23	24
21	4	5	6	65	21	24	25
22	4	5	6	66	22	24	25
23	4	6	7				
				67	22	25	26
24	5	6	7	68	22	25	26
25	5	7	7	69	23	25	27
26	6	7	8				
				70	23	26	27
27	6	7	8				
28	6	8	9	71	24	26	28
29	7	8	9	72	24	27	28
				73	25	27	28
30	7	9	10				
				74	25	28	29
31	7	9	10	75	25	28	29
32	8	9	10	76	26	28	30
33	8	10	11				
				77	26	29	30
34	9	10	11	78	27	29	31
35	9	11	12	79	27	30	31
36	9	11	12				
				80	28	30	32
37	10	12	13				
38	10	12	13	81	28	31	32
39	11	12	13	82	28	31	33
				83	29	32	33
40	11	13	14				
				84	29	32	33
41	11	13	14	85	30	32	34
42	12	14	15	86	30	33	34
43	12	14	15				
				87	31	33	35
44	13	15	16	88	31	34	35
45	13	15	16	89	31	34	36
46	13	15	16				
				90	32	35	36
47	14	16	17				
48	14	16	17				
49	15	17	18				

For values of $N > 90$, approximate values of r may be found by taking the nearest integer less than $\frac{1}{2}(N - 1) - k\sqrt{N + 1}$, where k is 1.29, 0.980, 0.822, for $P = 0.005, 0.025, 0.05$, respectively.

Reproduced in modified form with permission from W. J. Dixon and A. M. Mood, "The Statistical Sign Test," *Journal of the American Statistical Association*, Vol. 41 (1946), p. 560.

Table M—CRITICAL VALUES OF THE NUMBER OF RUNS, u

For a single sample, $N_1 = N_2$ equals the number of signs on each side of the median. P equals the probability that the number of runs, u, equals or $\begin{Bmatrix} \text{is less than} \\ \text{exceeds} \end{Bmatrix}$ the number listed under P in the $\begin{Bmatrix} \text{left} \\ \text{right} \end{Bmatrix}$ column.

P / $N_1 = N_2$	0.01	0.025	0.05	P / $N_1 = N_2$	0.01		0.025		0.05	
4	— —	2 —	2 —	25	17	34	18	33	19	32
5	— —	2 —	3 —	30	21	40	22	39	24	37
6	— —	3 —	3 —	35	25	46	27	44	28	43
7	— —	4 —	4 —	40	30	51	31	50	33	48
8	— —	4 —	5 —	45	34	57	36	55	37	54
9	— —	5 —	6 —							
10	— —	6 —	6 —	50	38	63	40	61	42	59
11	6 17	7 16	7 16	55	43	68	45	66	46	65
12	7 18	7 18	8 17	60	47	74	49	72	51	70
13	7 20	8 19	9 18	65	52	79	54	77	56	75
14	8 21	9 20	10 19	70	56	85	58	83	60	81
15	9 22	10 21	11 20	75	61	90	63	88	65	86
16	10 23	11 22	11 22	80	65	96	68	93	70	91
17	10 25	11 24	12 23	85	70	101	72	99	74	97
18	11 26	12 25	13 24	90	74	107	77	104	79	102
19	12 27	13 26	14 25	95	79	112	82	109	84	107
20	13 28	14 27	15 26	100	84	117	86	115	88	113

For $N_1 = N_2 > 100$, a normal curve may be used to approximate the probabilities with

$$\mu_u = \frac{2N_1N_2}{N_1 + N_2} + 1, \qquad \sigma_u^2 = \frac{2N_1N_2(2N_1N_2 - N_1 - N_2)}{(N_1 + N_2)^2(N_1 + N_2 - 1)}$$

Adapted with permission from F. Swed and C. Eisenhart, "Tables for Testing Randomness in Grouping in a Sequence of Alternatives," *Annals of Mathematical Statistics*, Vol. 14 (1943), pp. 66–87.

For two samples of N_1 and N_2 where $N_1 \leqq N_2$, P equals the probability that T' is equal to or $\begin{Bmatrix} \text{less} \\ \text{greater} \end{Bmatrix}$ than the numbers listed in the $\begin{Bmatrix} \text{left} \\ \text{right} \end{Bmatrix}$ columns.

N_1	N_2	P = 0.01		0.025		0.05	
2	5					3	13
2	6					3	15
2	7					3	17
2	8			3	19	4	18
2	9			3	21	4	20
2	10			3	23	4	22
3	3					6	15
3	4					6	18
3	5			6	21	7	20
3	6			7	23	8	22
3	7	6	27	7	26	8	25
3	8	6	30	8	28	9	27
3	9	7	32	8	31	10	29
3	10	7	35	9	33	10	32
4	4			10	26	11	25
4	5	10	30	11	29	12	28
4	6	11	33	12	32	13	31
4	7	11	37	13	35	14	34
4	8	12	40	14	38	15	37
4	9	13	43	15	41	16	40
4	10	13	47	15	45	17	43
5	5	16	39	17	38	19	36
5	6	17	43	18	42	20	40
5	7	18	47	20	45	21	44
5	8	19	51	21	49	23	47
5	9	20	55	22	53	24	51
5	10	21	59	23	57	26	54
6	6	24	54	26	52	28	50
6	7	25	59	27	57	29	55
6	8	27	63	29	61	31	59
6	9	28	68	31	65	33	63
6	10	29	73	32	70	35	67
7	7	34	71	36	69	39	66
7	8	36	76	38	74	41	71
7	9	37	82	40	79	43	76
7	10	39	87	42	84	45	81
8	8	46	90	49	87	51	85
8	9	48	96	51	93	54	90
8	10	50	102	53	99	56	96
9	9	59	112	63	108	66	105
9	10	61	119	65	115	69	111
10	10	74	136	78	132	82	128

For $N_1 > 10$ and $N_2 > 10$, $P(T' \leqq k)$, where k is an integer, can be approximated by means of a normal curve in x, with $x = k + \frac{1}{2}, \mu_x = \frac{1}{2}N_1(N_1 + N_2 + 1)$, and $\sigma_x^2 = \frac{1}{12}N_1 N_2(N_1 + N_2 + 1)$.

Table O—Critical Values Of Spearman's Rank Correlation Coefficient r_R.

For N paired variates, P is the approximate probability that a given value of r_R listed under it will be numerically attained or exceeded (two-tail probability).

P \ N	0.10	0.05	0.02	0.01
4	1.00			
5	.90	1.00	1.00	
6	.83	.87	.94	1.00
7	.71	.79	.89	.93
8	.64	.74	.83	.88
9	.60	.68	.78	.83
10	.56	.65	.75	.79
11	.52	.62	.74	.82
12	.50	.59	.70	.78
13	.48	.57	.67	.74
14	.46	.54	.64	.72
15	.44	.52	.62	.69
16	.42	.51	.60	.66
17	.41	.49	.58	.64
18	.40	.48	.56	.62
19	.39	.46	.55	.61
20	.38	.45	.53	.59
21	.37	.44	.52	.58
22	.36	.43	.51	.56
23	.35	.42	.50	.55
24	.34	.41	.48	.54
25	.34	.40	.48	.52
26	.33	.39	.46	.52
27	.32	.38	.46	.50
28	.32	.38	.45	.50
29	.31	.37	.44	.49
30	.31	.36	.43	.48

Adapted in part from tables in E. G. Olds, "Distributions of Sums of Squares of Rank Differences for Small Numbers of Individuals" and "The 5% Significance Levels for Sums of Squares of Rank Differences and a Correction," *Annals of Mathematical Statistics*, Vol. 9 (1938), pp. 133–148, and Vol. 20 (1949), pp. 117–118.

INDEX

313